Editorial Policy

§ 1. Lecture Notes aim to report new developments - quickly, informally, and at a high level. The texts should be reasonably self-contained and rounded off. Thus they may, and often will, present not only results of the author but also related work by other people. Furthermore, the manuscripts should provide sufficient motivation, examples and applications. This clearly distinguishes Lecture Notes manuscripts from journal articles which normally are very concise. Articles intended for a journal but too long to be accepted by most journals, usually do not have this "lecture notes" character. For similar reasons it is unusual for Ph. D. theses to be accepted for the Lecture Notes series.

§ 2. Manuscripts or plans for Lecture Notes volumes should be submitted (preferably in duplicate) either to one of the series editors or to Springer- Verlag, Heidelberg . These proposals are then refereed. A final decision concerning publication can only be made on the basis of the complete manuscript, but a preliminary decision can often be based on partial information: a fairly detailed outline describing the planned contents of each chapter, and an indication of the estimated length, a bibliography, and one or two sample chapters - or a first draft of the manuscript. The editors will try to make the preliminary decision as definite as they can on the basis of the available information.

§ 3. Final manuscripts should preferably be in English. They should contain at least 100 pages of scientific text and should include
- a table of contents;
- an informative introduction, perhaps with some historical remarks: it should be accessible to a reader not particularly familiar with the topic treated;
- a subject index: as a rule this is genuinely helpful for the reader.

Further remarks and relevant addresses at the back of this book.

Lecture Notes in Mathematics

1625

Editors:
A. Dold, Heidelberg
F. Takens, Groningen

Springer
Berlin
Heidelberg
New York
Barcelona
Budapest
Hong Kong
London
Milan
Paris
Santa Clara
Singapore
Tokyo

Serge Lang

Topics in Cohomology of Groups

Springer

Author

Serge Lang
Mathematics Derpartment
Yale University, Box 208 283
10 Hillhouse Avenue
New Haven, CT 06520-8283, USA

```
         Library of Congress Cataloging-in-Publication Data


Lang. Serge. 1927-
    [Rapport sur la cohomologie des groupes. English]
   Topics in cohomology of groups / Serge Lang.
        p.   cm. -- (Lecture notes in mathematics ; 1625)
   Includes bibliographical references (p.     -    ) and index.
   ISBN 3-540-61181-9 (alk. paper)
    1. Class field theory. 2. Group theory. 3. Homology theory.
 I. Title.  II. Series: Lecture notes in mathematics (Springer
 -Verlag) ; 1625.
 QA247.L3513   1996
 512'.74--dc20                                         96-26607
```

The first part of this book was originally published in French with the title
"Rapport sur la cohomologie des groupes" by Benjamin Inc., New York, 1996.
It was translated into English by the author for this edition. The last part
(pp. 188–215) is new to this edition.

Mathematics Subject Classification (1991): 11S25, 11S31, 20J06, 12G05, 12G10

ISBN 3-540-61181-9 Springer-Verlag Berlin Heidelberg New York

© Springer-Verlag Berlin Heidelberg 1996
Printed in Germany

Typesetting: Camera-ready T$_E$X output by the author
SPIN: 106 900 35   46/3111-54321-Printed on acid free paper

Contents

Chapter V. Augmented Products

Chapter VI. Spectral Sequences

Chapter VII. Groups of Galois Type (Unpublished article of Tate)

Chapter VIII. Group Extensions

Chapter IX. Class formations

Chapter X. Applications of Galois Cohomology in Algebraic Geometry (from letters of Tate)

Preface

The Benjamin notes which I published (in French) in 1966 on the cohomology of groups provided missing chapters to the Artin-Tate notes on class field theory, developed by cohomological methods. Both items were out of print for many years, but recently Addison-Wesley has again made available the Artin-Tate notes (which were in English). It seemed therefore appropriate to make my notes on cohomology again available, and I thank Springer-Verlag for publishing them (translated into English) in the Lecture Notes series.

The most basic necessary background on homological algebra is contained in the chapter devoted to this topic in my *Algebra* (derived functors and other material at this basic level). This material is partly based on what have now become routine constructions (Eilenberg-Cartan), and on Grothendieck's influential paper [Gr 59], which appropriately defined and emphasized δ-functors as such.

The main source for the present notes are Tate's private papers, and the unpublished first part of the Artin-Tate notes. The most significant exceptions are: Rim's proof of the Nakayama-Tate theorem, and the treatment of cup products, for which we have used the general notion of multilinear category due to Cartier.

The cohomological approach to class field theory was carried out in the late forties and early fifties, in Hochschild's papers [Ho 50a], [Ho 50b], [HoN 52], Nakayama [Na 41], [Na 52], Shafarevich [Sh 46], Weil's paper [We 51], giving rise to the Weil groups, and seminars of Artin-Tate in 1949-1951, published only years later [ArT 67].

As I stated in the preface to my *Algebraic Number Theory*, there are several approaches to class field theory. None of them makes any other obsolete, and each gives a different insight from the others.

The original Benjamin notes consisted of Chapters I through IX. Subsequently I wrote up Chapter X, which deals with applications to algebraic geometry. It is essentially a transcription of weekly installment letters which I received from Tate during 1958-1959. I take of course full responsibility for any errors which might have crept in, but I have made no effort to make the exposition anything more than a rough sketch of the material. Also the reader should not be surprised if some of the diagrams which have been qualified as being commutative actually have character -1.

The first nine chapters are basically elementary, depending only on standard homological algebra. The Artin-Tate axiomatization of class formations allows for an exposition of the basic properties of class field theory at this elementary level. Proofs that the axioms are satisfied are in the Artin-Tate notes, following Tate's article [Ta 52]. The material of Chapter X is of course at a different level, assuming some knowledge of algebraic geometry, especially some properties of abelian varieties.

I thank Springer Verlag for keeping all this material in print. I also thank Donna Belli and Mel Del Vecchio for setting the manuscript in AMSTeX, in a victory of person over machine.

Serge Lang
New Haven, 1995

CHAPTER I
Existence and Uniqueness

§1. The abstract uniqueness theorem

We suppose the reader is familiar with the terminology of abelian categories. However, we shall deal only with abelian categories which are categories of modules over some ring, or which are obtained from such in some standard ways, such as categories of complexes of modules. We also suppose that the reader is acquainted with the standard procedures constructing cohomological functors by means of resolutions with complexes, as done for instance in my *Algebra* (third edition, Chapter XX). In some cases, we shall summarize such constructions for the convenience of the reader.

Unless otherwise specified, all functors on abelian categories will be assumed additive. What we call a δ-**functor** (following Grothendieck) is sometimes called a **connected sequence of functors**. Such a functor is defined for a consecutive sequence of integers, and transforms an exact sequence

$$0 \to A \to B \to C \to 0$$

into an exact sequence

$$\cdots \to H^p(A) \to H^p(B) \to H^p(C) \xrightarrow{\delta} H^{p+1}(A) \to \cdots$$

functorially. If the functor is defined for all integers p with $-\infty < p < \infty$, then we say that this functor is **cohomological**.

Let H be a δ-functor on an abelian category $\mathfrak{A}$. We say that H is **erasable** by a subset $\mathfrak{M}$ of objects in A if for every A in $\mathfrak{A}$ there exists $M_A \in \mathfrak{M}$ and a monomorphism $\varepsilon_A : A \to M_A$ such that $H(M_A) = 0$. This definition is slightly more restrictive than the usual general definition (*Algebra*, Chapter XX, §7), but its conditions are those which are used in the forthcoming applications. An **erasing functor** for H consists of a functor

$$M : A \to M(A)$$

of $\mathfrak{A}$ into itself, and a monomorphism ε of the identity in M, i.e. for each object A we are given a monomorphism

$$\varepsilon_A : A \to M_A$$

such that, if $u : A \to B$ is a morphism in $\mathfrak{A}$, then there exists a morphism $M(u)$ and a commutative diagram

$$
\begin{array}{ccc}
0 \longrightarrow & A & \xrightarrow{\varepsilon_A} M(A) \\
& u \downarrow & \downarrow M(u) \\
0 \longrightarrow & B & \xrightarrow[\varepsilon_B]{} M(B)
\end{array}
$$

such that $M(uv) = M(u)M(v)$ for the composite of two morphisms u, v. In addition, one requires $H(M_A) = 0$ for all $A \in \mathfrak{A}$.

Let $X(A) = X_A$ be the cokernel of ε_A. For each u there is a morphism

$$X(u) : X_A \to X_B$$

such that the following diagram is commutative:

$$
\begin{array}{ccccccc}
0 \longrightarrow & A & \longrightarrow & M_A & \longrightarrow & X_A & \longrightarrow 0 \\
& u \downarrow & & \downarrow M(u) & & \downarrow X(u) & \\
0 \longrightarrow & B & \longrightarrow & M_B & \longrightarrow & X_B & \longrightarrow 0,
\end{array}
$$

and for the composite of two morphisms u, v we have $X(uv) = X(u)X(v)$. We then call X the **cofunctor** of M.

Let p_0 be an integer, and $H = (H^p)$ a δ-functor defined for some values of p. We say that M is an **erasing functor for H in dimension $> p_0$** if $H^p(M_A) = 0$ for all $A \in \mathfrak{A}$ and all $p > p_0$.

We have similar notions on the left. Let H be an exact δ-functor on $\mathfrak{A}$. We say that H is **coerasable** by a subset $\mathfrak{M}$ if for each object A there exists an epimorphism

$$\eta_A : M_A \to A$$

with $M_A \in \mathfrak{M}$, such that $H(M_A) = 0$. A **coerasing functor** M for H consists of an epimorphism of M with the identity. If η is such a functor, and $u : A \to B$ is a morphism, then we have a commutative diagram with exact horizontal sequences:

$$
\begin{array}{ccccccccc}
0 & \longrightarrow & Y_A & \longrightarrow & M_A & \xrightarrow{\eta_A} & A & \longrightarrow & 0 \\
& & \Big\downarrow{\scriptstyle Y(u)} & & \Big\downarrow{\scriptstyle M(u)} & & \Big\downarrow{\scriptstyle u} & & \\
0 & \longrightarrow & Y_B & \longrightarrow & M_B & \xrightarrow[\eta_B]{} & B & \longrightarrow & 0
\end{array}
$$

and Y_A is functorial in A, i.e. $Y(uv) = Y(u)Y(v)$.

Remark. In what follows, erasing functors will have the additional property that the exact sequence associated with each object A will split over $\mathbf{Z}$, and therefore remains exact under tensor products or hom. An erasing functor into an abelian category of abelian groups having this property will be said to be **splitting**.

Theorem 1.1. First uniqueness theorem. *Let $\mathfrak{A}$ be an abelian category. Let H, F be two δ-functors defined in degrees $0, 1$ (resp. $0, -1$) with values in the same abelian category. Let (φ_0, φ_1) and $(\varphi_0, \bar{\varphi}_1)$ be δ-morphisms of H into F, coinciding in dimension 0 (resp. $(\varphi_{-1}, \varphi_0)$ and $(\bar{\varphi}_{-1}, \varphi_0)$). Suppose that H^1 is erasable (resp. H^{-1} is coerasable). Then we have $\varphi_1 = \bar{\varphi}_1$ (resp. $\varphi_{-1} = \bar{\varphi}_{-1}$).*

Proof. The proof being self dual, we give it only for the case of indices $(0, 1)$. For each object $A \in \mathfrak{A}$ we have an exact sequence

$$0 \to A \to M_A \to X_A \to 0$$

and $H^1(M_A) = 0$. There is a commutative diagram

$$
\begin{array}{ccccccc}
H^0(M_A) & \longrightarrow & H^0(X_A) & \xrightarrow{\delta_H} & H^1(A) & \longrightarrow & 0 \\
\Big\downarrow{\scriptstyle \varphi_0} & & \Big\downarrow{\scriptstyle \varphi_0} & & \Big\downarrow{\scriptstyle \varphi_1, \bar{\varphi}_1} & & \\
F^0(M_A) & \longrightarrow & F^0(X_A) & \xrightarrow{\delta_F} & F^1(A) & \longrightarrow & 0
\end{array}
$$

with horizontal exact sequences, from which it follows that δ_H is surjective. It follows at once that $\varphi_1 = \bar{\varphi}_1$.

In the preceding theorem, φ_1 and $\bar{\varphi}_1$ are given. One can also prove a result which implies their existence.

Theorem 1.2. Second uniqueness theorem. *Let $\mathfrak{A}$ be an abelian category. Let H, F be δ-functors defined in degrees $(0,1)$ (resp. $0, -1$) with values in the same abelian category. Let $\varphi_0 : H^0 \xrightarrow{\cdot} F^0$ be a morphism. Suppose that H^1 is erasable by injectives (resp. H^{-1} is coerasable by projectives). Then there exists a unique morphism*

$$\varphi_1 : H^1 \to F^1 \ (\text{resp. } \varphi_{-1} : H^{-1} \to F^{-1})$$

such that (φ_0, φ_1) (resp. $(\varphi_0, \varphi_{-1})$) is also a δ-morphism. The association $\varphi_0 \mapsto \varphi_1$ is functorial in a sense made explicit below.

Proof. Again the proof is self dual and we give it only in the cases when the indices are $(0,1)$. For each object $A \in \mathfrak{A}$ we have the exact sequence

$$0 \to A \to M_A \to X_A \to 0$$

and $H^1(M_A) = 0$. We have to define a morphism

$$\varphi_1(A) : H^1(A) \to F^1(A)$$

which commutes with the induced morphisms and with δ. We have a commutative diagram

$$
\begin{array}{ccccccc}
H^0(M_A) & \longrightarrow & H^0(X_A) & \xrightarrow{\delta_H} & H^1(A) & \longrightarrow & 0 \\
\varphi_0 \downarrow & & \varphi_0 \downarrow & & & & \\
F^0(M_A) & \longrightarrow & F^0(X_A) & \xrightarrow[\delta_F]{} & F^1(A) & &
\end{array}
$$

with exact horizontal sequences. The right surjectivity is just the erasing hypothesis. The left square commutativity shows that $\mathrm{Ker}\ \delta_H$ is contained in the kernel of $\delta_F \varphi_0(X_A)$. Hence there exists a unique morphism

$$\varphi_1(A) : H^1(A) \to F^1(A)$$

which makes the right square commutative. We shall prove that $\varphi_1(A)$ satisfies the desired conditions.

First, let $u : A \to B$ be a morphism. From the hypotheses, there exists a commutative diagram

$$
\begin{array}{ccccccccc}
0 & \longrightarrow & A & \longrightarrow & M_A & \longrightarrow & X_A & \longrightarrow & 0 \\
& & \downarrow{\scriptstyle u} & & \downarrow{\scriptstyle M(u)} & & \downarrow{\scriptstyle X(u)} & & \\
0 & \longrightarrow & B & \longrightarrow & M_B & \longrightarrow & X_B & \longrightarrow & 0
\end{array}
$$

the morphism $M(u)$ being defined because M_A is injective. The morphism $X(u)$ is then defined by making the right square commutative. To simplify notation, we shall write u instead of $M(u)$ and $X(u)$.

We consider the cube:

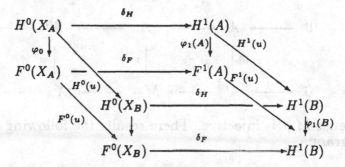

We have to show that the right face is commutative. We have:

$$
\begin{aligned}
\varphi_1(B)H^1(u)\delta_H &= \varphi_1(B)\delta_H H^0(u) \\
&= \delta_F \varphi_0 H^0(u) \\
&= \delta_F F^0(u)\varphi_0 \\
&= F^1(u)\delta_F \varphi_0 \\
&= F^1(u)\varphi_1(A)\delta_H .
\end{aligned}
$$

We have used the fact (implied by the hypotheses) that all the faces of the cube are commutative except possibly the right face. Since δ_H is surjective, one gets what we want, namely

$$
\varphi_1(B)H^1(u) = F^1(u)\varphi_1(A).
$$

The above argument may be expressed in the form of a useful general lemma.

If, in a cube, all the faces are commutative except possibly one, and one of the arrows as above is surjective, then this face is also commutative.

Next we have to show that φ_1 commutes with δ, that is (φ_0, φ_1) is a δ-morphism. Let

$$0 \to A' \to A \to A'' \to 0$$

be an exact sequence in $\mathfrak{A}$. Then there exist morphisms

$$v : A \to M_{A'} \qquad \text{and} \qquad w : A'' \to X_{A'}$$

making the following diagram commutative:

$$
\begin{array}{ccccccccc}
0 & \longrightarrow & A' & \longrightarrow & A & \longrightarrow & A'' & \longrightarrow & 0 \\
& & \downarrow {\scriptstyle id} & & \downarrow {\scriptstyle v} & & \downarrow {\scriptstyle w} & & \\
0 & \longrightarrow & A' & \longrightarrow & M_{A'} & \longrightarrow & X_{A'} & \longrightarrow & 0
\end{array}
$$

because $M_{A'}$ is injective. There results the following commutative diagram:

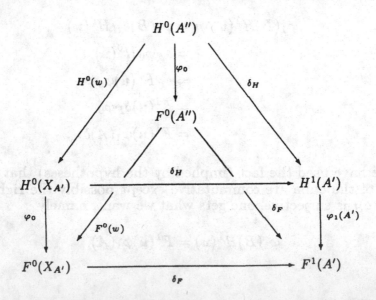

We have to show that the right square is commutative. Note that the top and bottom triangles are commutative by definition of a δ-functor. The left square is commutative by the hypothesis that φ_0 is a morphism of functors. The front square is commutative by definition of $\varphi_1(A')$. We thus find

$$
\begin{aligned}
\varphi_1(A')\delta_H &= \varphi_1(A')\delta_H H^0(w) &&\text{(top triangle)} \\
&= \delta_F \varphi_0 H^0(w) &&\text{(front square)} \\
&= \delta_F F^0(w)\varphi_0 &&\text{(left square)} \\
&= \delta_F \varphi_0 &&\text{(bottom triangle),}
\end{aligned}
$$

which concludes the proof.

Finally, let us make explicit what we mean by saying that φ_1 depends functorially on φ_0. Suppose we have three functors H, F, E defined in degrees $0, 1$; and suppose given $\varphi_0 : H^0 \to F^0$ and $\psi_0 : F^0 \to E^0$. Suppose in addition that the erasing functor erases both H^1 and F^1. We can then construct φ_1 and ψ_1 by applying the theorem. On the other hand, the composite

$$
\psi_0 \varphi_0 = \theta_0 : H^0 \to E^0
$$

is also a morphism, and the theorem implies the existence of a morphism

$$
\theta_1 : H^1 \to E^1
$$

such that (θ_0, θ_1) is a δ-morphism. By uniqueness, we obtain

$$
\theta_1 = \psi_1 \circ \varphi_1.
$$

This is what we mean by the assertion that φ_1 depends functorially on φ_0.

§2. Notation, and the uniqueness theorem in Mod(G)

We now come to the cohomology of groups. Let G be a group. As usual, we let $\mathbf{Q}$ and $\mathbf{Z}$ denote the rational numbers and the integers respectively. Let $\mathbf{Z}[G]$ be the group ring over $\mathbf{Z}$. Then

$\mathbf{Z}[G]$ is a free module over $\mathbf{Z}$, the group elements forming a basis over $\mathbf{Z}$. Multiplicatively, we have

$$\left(\sum_{\sigma \in G} a_\sigma \sigma\right)\left(\sum_{\tau \in G} b_\tau \tau\right) = \sum_{\sigma, \tau} a_\sigma b_\tau \sigma \tau,$$

the sums being taken over all elements of G, but only a finite number of a_σ and b_τ being $\neq 0$. Similarly, one defines the group algebra $k[G]$ over an arbitrary commutative ring k.

The group ring is often denoted by $\Gamma = \Gamma_G$. It contains the ideal I_G which is the kernel of the **augmentation homomorphism**

$$\varepsilon : \mathbf{Z}[G] \to \mathbf{Z}$$

defined by $\varepsilon\left(\sum n_\sigma \sigma\right) = \sum n_\sigma$. One sees at once that I_G is $\mathbf{Z}$-free, with a basis consisting of all elements $\sigma - e$, with σ ranging over the elements of G not equal to the unit element. Indeed, if $\sum n_\sigma = 0$, then we may write

$$\sum n_\sigma \sigma = \sum n_\sigma (\sigma - e).$$

Thus we obtain an exact sequence

$$0 \to I_G \to \mathbf{Z}[G] \to \mathbf{Z} \to 0,$$

used constantly in the sequel. The sequence splits, because $\mathbf{Z}[G]$ is a direct sum of I_G and $\mathbf{Z} \cdot e_G$ (identified with $\mathbf{Z}$).

Abelian groups form an abelian category, equal to the category of $\mathbf{Z}$-modules, denoted by $\mathrm{Mod}(\mathbf{Z})$. Similarly, the category of modules over a ring R will be denoted by $\mathrm{Mod}(R)$.

An abelian group A is said to be a G-**module** if one is given an operation (or action) of G on A; in other words, one is given a map

$$G \times A \to A$$

satisfying

$$(\sigma \tau)a = \sigma(\tau a) \qquad e \cdot a = a \qquad \sigma(a + b) = \sigma a + \sigma b$$

for all $\sigma, \tau \in G$ and $a, b \in A$. We let $e = e_G$ be the unit element of G. One extends this operation by linearity to the group ring $\mathbf{Z}[G]$. Similarly, if k is a commutative ring and A is a k-module, one extends the operation of G on A to $k[G]$ whenever the operation of G commutes with the operation of k on A. Then the category of $k[G]$-modules is denoted by $\text{Mod}_k(G)$ or $\text{Mod}(k, G)$.

The G-modules form an abelian category, the morphisms being the G-homomorphisms. More precisely, if $f : A \to B$ is a morphism in $\text{Mod}(\mathbf{Z})$, and if A, B are also G-modules, then G operates on $\text{Hom}(A, B)$ by the formula

$$(\sigma f)(a) = \sigma(f(\sigma^{-1}a)) \quad \text{for} \quad a \in A \quad \text{and} \quad \sigma \in G.$$

If there is any danger of confusion one may write $[\sigma]f$ to denote this operation. If $[\sigma]f = f$, one says that f is a G-**homomorphism**, or a G-**morphism**. The set of G-morphisms from A into B is an abelian group denoted by $\text{Hom}_G(A, B)$. The category consisting of G-modules and G-morphisms is denoted by $\text{Mod}(G)$. It is the same as $\text{Mod}(\Gamma_G)$.

Let $A \in \text{Mod}(G)$. We let A^G denote the submodule of A consisting of all elements $a \in A$ such that $\sigma a = a$ for all $\sigma \in G$. In other words, it is the submodule of fixed elements by G. Then A^G is an abelian group, and the association

$$H_G^0 : A \mapsto A^G$$

is a functor from $\text{Mod}(G)$ into the category of abelian groups, also denoted by **Grab**. This functor is left exact.

We let $\varkappa_G$ denote the canonical map (in the present case the identity) of an element $a \in A^G$ into $H_G^0(A)$.

Theorem 2.1. *Let H_G be a cohomological functor on $\text{Mod}(G)$ with values in $\text{Mod}(\mathbf{Z})$, and such that H_G^0 is defined as above. Assume that $H_G^r(M) = 0$ if M is injective and $r > 1$. Assume also that $H_G^r(A) = 0$ for $A \in \text{Mod}(G)$ and $r < 0$. Then two such cohomological functors are isomorphic, by a unique morphism which is the identity on $H_G^0(A)$.*

This theorem is just a special case of the general uniqueness theorem.

Corollary 2.2. *If $G = \{e\}$ then $H_G^r(A) = 0$ for all $r > 0$.*

Proof. Define H_G by letting $H_G^0(A) = A^G$ and $H_G^r(A) = 0$ for $r \neq 0$. Then it is immediately verified that H_G is a cohomological functor, to which we can apply the uniqueness theorem.

Corollary 2.3. *Let $n \in \mathbf{Z}$ and let $n_A : A \to A$ be the morphism $a \mapsto na$ for $a \in A$. Then $H_G^r(n_A) = n_H$ (where H stands for $H_G^r(A)$).*

Proof. Since the coboundary δ is additive, it commutes with multiplication by n, and again we can apply the uniqueness theorem.

The existence of the functor H_G will be proved in the next section.

We say that G **operates trivially** on A if $A = A^G$, that is $\sigma a = a$ for all $a \in A$ and $\sigma \in G$. We always assume that G operates trivially on $\mathbf{Z}, \mathbf{Q},$ and $\mathbf{Q}/\mathbf{Z}$.

We define the abelian group

$$A_G = A/IA_G.$$

This is the factor group of A by the subgroup of elements of the form $(\sigma - e)a$ with $\sigma \in G$ and $a \in A$. The association

$$A \mapsto A_G$$

is a functor from $\text{Mod}(G)$ into Grab.

Let U be a subgroup of finite index in G. We may then define the **trace**

$$S_G^U : A^U \to A^G \qquad \text{by the formula} \qquad S_G^U(a) = \sum_c \bar{c}a,$$

where $\{c\}$ is the set of left cosets of U in G, and $\bar{c}$ is a representative of c, so that

$$G = \bigcup_c \bar{c}U.$$

If $U = \{e\}$, then G is finite, and in that case the **trace** is written S_G, so

$$S_G(a) = \sum_{\sigma \in G} \sigma a.$$

For the record, we state the following useful lemma.

Lemma 2.4. *Let A, B, C be G-modules. Let U be a subgroup of finite index in G. Let*

$$A \xrightarrow{u} B \xrightarrow{v} C \xrightarrow{w} D$$

be morphisms in $\mathrm{Mod}(G)$, nad suppose that u, w are G-morphisms while v is a U-morphism. Then

$$\mathbf{S}_G^U(wvu) = w\mathbf{S}_G^U(v)u.$$

Proof. Immediate.

We shall now describe some embedding functors in $\mathrm{Mod}(G)$. These will turn out to erase some cohomological functors to be defined later. Indeed, injective or projective modules will not suffice to erase cohomology, for several reasons. First, when we change the group G, an injective does not necessarily remain injective. Second, an exact sequence

$$0 \to A \to J \to A'' \to 0$$

with an injective module J does not necessarily remain exact when we take its tensor product with an arbitrary module B. Hence we shall consider another class of modules which behave better in both respects.

Let G be a group and let B be an abelian group, i.e. a $\mathbb{Z}$-module. We denote by $M_G(B)$ or $M(G, B)$ the set of functions from G into B, these forming an abelian group in the usual way (adding the values). We make $M_G(B)$ into a G-module by defining an operation of G by the formulas

$$(\sigma f)(x) = f(x\sigma), \qquad \text{for} \qquad x, \sigma \in G.$$

We have trivially $(\sigma\tau)(f) = \sigma(\tau f)$. Furthermore:

Proposition 2.6. *Let G' be a subgroup of G and let $G = \bigcup_\alpha x_\alpha G'$ be a coset decomposition. For $f \in M(G, B)$ let f_α be the function in $M(G', B)$ such that $f_\alpha(y) = f(x_\alpha y)$ for $y \in G'$. Then the map*

$$f \mapsto \prod_\alpha f_\alpha$$

is an isomorphism

$$M(G, B) \xrightarrow{\approx} \prod_\alpha M(G', B)$$

in the category of G'-modules.

The proof is immediate, and Proposition 2.5 is a special case with G' equal to the trivial subgroup.

Let $A \in \mathrm{Mod}(G)$, and define

$$\varepsilon_A : A \to M_G(A)$$

by the condition that $\varepsilon_A(a)$ is the function f_a such that $f_a(\sigma) = \sigma a$ for all $a \in A$ and $\sigma \in G$. We then obtain an exact sequence

(1) $$0 \to A \xrightarrow{\varepsilon_A} M_G(A) \to X_A \to 0$$

in $\mathrm{Mod}(G)$. Furthermore, this sequence splits over $\mathbf{Z}$, because the map

$$M_G(A) \to A \qquad \text{given by} \qquad f \mapsto f(e)$$

splits the left arrow in this sequence, i.e. composed with ε_A it yields the identity on A. Consequently tensoring this sequence with an arbitrary G-module B preserves exactness.

We already know that M_G is an exact functor. In addition, if $f : A \to B$ is a morphism in $\mathrm{Mod}(G)$, then in the following diagram

(2)
$$\begin{array}{ccccccccc}
0 & \longrightarrow & A & \xrightarrow{\varepsilon_A} & M_G(A) & \longrightarrow & X_A & \longrightarrow & 0 \\
& & \downarrow{\scriptstyle f} & & \downarrow{\scriptstyle M_G(f)} & & \downarrow{\scriptstyle X(f)} & & \\
0 & \longrightarrow & B & \xrightarrow[\varepsilon_B]{} & M_G(B) & \longrightarrow & X_B & \longrightarrow & 0
\end{array}$$

the left square is commutative, and hence the right square is commutative. Therefore, we find:

Theorem 2.5. *Let G be a group. Notations as above, the pair (M_G, ε) is an embedding functor in $\mathrm{Mod}(G)$. The associated exact sequence (1) splits over $\mathbf{Z}$ for each $A \in \mathrm{Mod}(G)$.*

In the next section, we shall define a cohomological functor H_G on $\mathrm{Mod}(G)$ for which (M_G, ε) is an erasing functor. By Proposition 2.6, we shall then find:

Corollary 2.6. *Let G' be a subgroup of G, and consider $\mathrm{Mod}(G)$ as a subcategory of $\mathrm{Mod}(G')$. Then $H_{G'}$ is a cohomological functor on $\mathrm{Mod}(G)$, and (M_G, ε) is an erasing functor for $H_{G'}$.*

Thus we shall have achieved our objective of finding a serviceable erasing functor simultaneously for a group and its subgroups, behaving properly under tensor products. The erasing functor as above will be called the **ordinary erasing functor**.

Remark. Let U be a subgroup of finite index in G. Let $A, B \in \mathrm{Mod}(G)$. Let $f : A \to B$ be a U-morphism. We may take the trace

$$\mathbf{S}_G^U(f) : A \to B$$

which is a G-morphism. Furthermore, considering $\mathrm{Mod}(G)$ as a subcategory of $\mathrm{Mod}(U)$, we see that (M_G, ε) is an embedding functor relative to U, that is, there exist U-morphisms $M_G(f)$ and $X(f)$ such that the diagram (2) is commutative, but with vertical U-morphisms.

Applying the trace to these vertical morphisms, and using Lemma 2.4, we obtain a commutative diagram:

$$
\begin{array}{ccccccccc}
0 & \longrightarrow & A & \xrightarrow{\varepsilon_A} & M_G(A) & \longrightarrow & X_A & \longrightarrow & 0 \\
 & & {\scriptstyle \mathbf{S}_G^U(f)}\downarrow & & {\scriptstyle \mathbf{S}_G^U M_G(f)}\downarrow & & {\scriptstyle \mathbf{S}_G^U X(f)}\downarrow & & \\
0 & \longrightarrow & B & \xrightarrow[\varepsilon_B]{} & M_G(B) & \longrightarrow & X_B & \longrightarrow & 0
\end{array}
$$

(3)

The case of finite groups

For the rest of this section, we assume that G is finite.

We define two functors from $\mathrm{Mod}(G)$ into $\mathrm{Mod}(\mathbf{Z})$ by

$$\mathbf{H}_G^0 : A \mapsto A^G/\mathbf{S}_G A$$
$$\mathbf{H}_G^{-1} : A \mapsto A_{\mathbf{S}_G}/I_G A.$$

We denote by $A_{\mathbf{S}_G}$ the kernel of $\mathbf{S}_G$ in A. This is a special case of the notation whereby if $f : A \to B$ is a homomorphism, we let A_f be its kernel.

We let

$$\varkappa_G : A^G \to \mathbf{H}_G^0(A) = A^G/\mathbf{S}_G A.$$

$$\varkappa_G : A_{\mathbf{S}_G} \to \mathbf{H}_G^{-1}(A) = A_{\mathbf{S}_G}/I_G A.$$

be the canonical maps. The proof of the following result is easy and straightforward, and will be left to the reader.

Theorem 2.7. *The functors* $\mathbf{H}_G^{-1}$ *and* $\mathbf{H}_G^0$ *form a* δ-*functor if one defines the coboundary as follows. Let*

$$0 \to A' \xrightarrow{u} A \xrightarrow{v} A'' \to 0$$

be an exact sequence in $\mathrm{Mod}(G)$. *For* $a'' \in A''_{\mathbf{S}_G}$ *we define*

$$\delta \varkappa_G(a'') = \varkappa_G(u^{-1}\mathbf{S}_G v^{-1} a'').$$

The inverse images in this last formula have the usual meaning. One chooses any element a such that $va = a''$, then one takes the trace $\mathbf{S}_G$. One shows that this is an element in the image of u, so we can take u^{-1} of this element to be an element of A'^G, whose class modulo $\mathbf{S}_G A'$ is well defined, i.e. is independent of the choices of a such that $va = a''$. The verification of these assertions is trivial, and left to the reader. (Cf. *Algebra*, Chapter III, §9.)

Theorem 2.8. *Let* $\mathbf{H}_G$ *be a cohomological functor on* $\mathrm{Mod}(G)$ *(with finite group* G), *with value in* $\mathrm{Mod}(\mathbf{Z})$, *and such that* $\mathbf{H}_G^0$ *is as above. Suppose that* $\mathbf{H}_G^r(M) = 0$ *if* M *is injective and* $r \leqq 0$. *If* $\mathbf{F}_G$ *is another cohomological functor having the same properties, then there exists a unique isomorphism of* $\mathbf{H}_G$ *with* $\mathbf{F}_G$ *which is the identity on* $\mathbf{H}_G^0$.

Proof. This is a particular case of the uniqueness theorem.

Corollary 1.9. *If* G *is trivial then* $\mathbf{H}_G^r(A) = 0$ *for all* $r \in \mathbf{Z}$.

Corollary 2.10. *Let* $n \in \mathbf{Z}$ *and suppose* G *finite. For* A *in* $\mathrm{Mod}(G)$ *we have* $\mathbf{H}_G^r(n_A) = n_{\mathbf{H}}$ *(abbreviating* $\mathbf{H}_G^r$ *by* $\mathbf{H}$).

Both corollaries are direct consequences of the uniqueness theorem, like their counterpart for the other functor, as in Corollaries 2.2 and 2.3.

Let K_1, K_2, K be commutative rings, and let T be a biadditive bifunctor

$$T : \mathrm{Mod}(K_1) \times \mathrm{Mod}(K_2) \to \mathrm{Mod}(K).$$

Suppose we are given an action of G on $A_1 \in \mathrm{Mod}(K_1)$ and on $A_2 \in \mathrm{Mod}(K_2)$, so $A_1 \in \mathrm{Mod}(K_1[G])$ and $A_2 \in \mathrm{Mod}(K_2[G])$. Then $T(A_1, A_2)$ is a $K[G]$-module, under the operation $T(\sigma, \sigma)$ if T is covariant in both variables, and $T(\sigma^{-1}, \sigma)$ if T is contravariant in the first variable and covariant in the second. This remark will be applied to the case when T is the tensor product or $T = \mathrm{Hom}$.

A $K[G]$-module A is called $K[G]$-**regular** if the identity 1_A is a trace, that is there exists a K-morphism $v : A \to A$ such that

$$1_A = \mathbf{S}_G(v).$$

When $K = \mathbf{Z}$, a $K[G]$-regular module is simply called G-**regular**.

Proposition 2.11. *Let K_1, K_2, K be commutative rings as above, and let T be as above. Let $A_i \in \mathrm{Mod}(K_i)$ $(i = 1, 2)$ and suppose A_i is $K_i[G]$-regular for $i = 1, 2$. Then $T(A_1, A_2)$ is $K[G]$-regular.*

Proof. Left to the reader.

Proposition 2.12. *Let G' be a subgroup of the finite group G. Let $A \in \mathrm{Mod}(K, G)$. If A is $K[G]$-regular then A is also $K[G']$-regular. If G' is normal in G, then $A^{G'}$ is $K[G/G']$-regular.*

Proof. Write $G = \bigcup G' x_i$ expressing G as a right coset decomposition of G'. Then by assumption, we can write 1_A in the form

$$1_A = \sum_{\tau \in G'} \sum_i \tau x_i v$$

with some K-morphism v. The two assertions of the proposition are then clear, according as we take the double sum in the given order, or reverse the order of the summation.

Proposition 2.13. *Let $A \in \mathrm{Mod}(K, G)$. Then A is $K[G]$-projective if and only if A is K-projective and $K[G]$-regular.*

Proof. We recall that a projective module is characterized by being a direct summand of a free module. Suppose that A is $K[G]$-projective. We may then write A as a direct summand of a free

$K[G]$-module $F = A \oplus B$, with the natural injection i and projection π in the sequence

$$A \xrightarrow{i} A \times B = F \xrightarrow{\pi} A$$

with $\pi i = 1_A$, and both i, π are $K[G]$-homomorphisms. Since F is $K[G]$-free, it follows that $1_F = \mathbf{S}_G(v)$ for some K-homomorphism v. Then by Lemma 2.4,

$$1_A = \pi 1_F i = \pi \mathbf{S}_G(v) i = \mathbf{S}_G(\pi v i),$$

whence A is $K[G]$-regular. Conversely, let A be K-projective and $K[G]$-regular. Let

$$F \xrightarrow{\pi} A \to 0$$

be an exact sequence in $\mathrm{Mod}(K, G)$, with F being $K[G]$-free. By hypothesis, there exists a K-morphism $i_K : A \to F$ such that $\pi i_K = 1_A$, and there exists a K-morphism $v : A \to A$ such that $1_A = \mathbf{S}_G(v)$. We then find

$$\pi \mathbf{S}_G(i_K v) = \mathbf{S}_G(\pi i_K v) = \mathbf{S}_G(v) = 1_A,$$

which shows that $\mathbf{S}_G(i_K v)$ splits π, whence A is a direct summand of a free module, and is therefore $K[G]$-projective. This proves the proposition.

With the same type of proof, taking the trace of a projection, one also obtains the following result.

Proposition 2.14. *In* $\mathrm{Mod}(G)$, *a direct summand of a G-regular module is also G-regular. In particular, every projective module in* $\mathrm{Mod}(G)$ *is also G-regular.*

Proof. The second assertion is obvious for free modules, whence it follows from the first assertion for projectives.

For finite groups we have a modification of the embedding functor defined previously for arbitrary groups, and this modification will enjoy stronger properties. We consider the following two exact sequences:

(3) $$0 \to I_G \to \mathbf{Z}[G] \xrightarrow{\varepsilon} \mathbf{Z} \to 0$$

(4) $$0 \to \mathbf{Z} \xrightarrow[\varepsilon']{} \mathbf{Z}[G] \to J_G \to 0.$$

The first one is just the one already considered, with the augmentation homomorphism ε. The second is defined as follows. We embed $\mathbf{Z}$ in $\mathbf{Z}[G]$ on the diagonal, that is

$$\varepsilon' : n \mapsto n \sum_{\sigma \in G} \sigma.$$

Since G acts trivially on $\mathbf{Z}$, it follows that ε' is a G-homomorphism. We denote its cokernel by J_G.

Proposition 2.15. *The exact sequences* (3) *and* (4) *split in* Mod($\mathbf{Z}$).

Proof. We already know this for (3). For (4), given $\xi = \sum n_\sigma \sigma$ in $\mathbf{Z}[G]$ we have a decomposition

$$\xi = n_e \left(\sum_{\sigma \in G} \sigma \right) + \sum_{\sigma \neq e} (n_\sigma - n_e) \sigma$$

$$\in \mathbf{Z} \left(\sum_{\sigma \in G} \sigma \right) + \sum_{\sigma \neq e} \mathbf{Z} \sigma,$$

which shows that $\mathbf{Z}[G]$ is a direct sum of $\varepsilon'(\mathbf{Z})$ and another module, as was to be shown.

Given any $A \in \text{Mod}(G)$, taking the tensor product (over $\mathbf{Z}$) of the split exact sequences (3) and (4) with A yields split exact sequences by a basic elementary property of the tensor product, with G-morphisms $\varepsilon_A = \varepsilon \otimes 1_A$ and $\varepsilon'_A = \varepsilon' \otimes 1_A$, as shown below:

(4A) $0 \to I_G \otimes A \to \mathbf{Z}[G] \otimes A \xrightarrow{\varepsilon_A} \mathbf{Z} \otimes A = A \to 0$

(5A) $0 \to A = \mathbf{Z} \otimes A \underset{\varepsilon'_A}{\longrightarrow} \mathbf{Z}[G] \otimes A \to J_G \otimes A \to 0.$

As usual, we identify $\mathbf{Z} \otimes A$ with A. Let $\mathbf{M}_G$ be the functor given by

$$\mathbf{M}_G(A) = \mathbf{Z}[G] \otimes A.$$

We observe that $\mathbf{M}_G(A)$ is G-regular. In the next section, we shall define a cohomological functor on $\text{Mod}(G)$ for which $\mathbf{M}_G$ will be an erasing functor.

Let $f : A \to B$ be a G-morphism, or more generally, suppose G' is a subgroup of G and $A, B \in \operatorname{Mod}(G)$, while f is a G'-morphism. Then

$$\mathbf{M}_G(f) = 1 \otimes f$$

is a G'-morphism.

§3. Existence of the cohomological functors on $\operatorname{Mod}(G)$

Although we reproduced the proofs of uniqueness, because they were short, we now assume that the reader is acquainted with standard facts of general homology theory. These are treated in *Algebra*, Chapter XX, of which we now use §8, especially Proposition 8.2 giving the existence of the derived functors. We apply this proposition to the bifunctor

$$T(A, B) = \operatorname{Hom}_G(A, B) \quad \text{for} \quad A, B \in \operatorname{Mod}(G),$$

with an arbitrary group G. We have

$$\operatorname{Hom}_G(\mathbf{Z}, A) = A^G.$$

We then find:

Theorem 3.1. *Let X be a projective resolution of $\mathbf{Z}$ in $\operatorname{Mod}(G)$. Let $H(A)$ be the homology of the complex $\operatorname{Hom}_G(X, A)$. Then $H = \{H^r\}$ is a cohomology functor on $\operatorname{Mod}(G)$, such that*

$$H^r(A) = 0 \text{ if } r < 0.$$
$$H^0(A) = A^G.$$
$$H^r(A) = 0 \text{ if } A \text{ is injective in } \operatorname{Mod}(G) \text{ and } r \geq 1.$$

This cohomology functor is determined up to a unique isomorphism.

For the convenience of the reader, we write the first few terms of the sequences implicit in Theorem 3.1. From the resolution

$$\cdots \to X_1 \to X_0 \to \mathbf{Z} \to 0$$

we obtain the sequence

$$0 \to \operatorname{Hom}_G(\mathbf{Z}, A) \to \operatorname{Hom}_G(X_0, A) \to \operatorname{Hom}_G(X_1, A) \to$$

so the cohomology sequence arising from an exact sequence

$$0 \to A' \to A \to A'' \to 0$$

starts with an exact part

$$0 \to A'^G \to A^G \to A''^G \to H^1(A').$$

Dually, we work with the tensor product. We let I_G as before be the augmentation ideal. For $A \in \mathrm{Mod}(G)$, $I_G A$ is the G-module generated by all elements $\sigma a - a$ with $a \in A$, and even consists of such elements. We consider the functor $A \mapsto A_G = A/I_G A$ from $\mathrm{Mod}(G)$ into Grab. For $A, B \in \mathrm{Mod}(G)$ we define

$$T_G(A, B) = A \otimes_G B = (A \otimes B)_G.$$

Then T_G is a bifunctor

$$T_G : \mathrm{Mod}(G) \times \mathrm{Mod}(G) \to \mathrm{Grab},$$

covariant in both variables. From *Algebra*, Chapter XX, Proposition 3.2', we find:

Theorem 3.2. *Let X be a projective resolution of $\mathbf{Z}$ in $\mathrm{Mod}(G)$. Let T_G be as above, and let $H = \{H_r\}$ be the homology of the complex $T_G(X, A)$. Then H is a homological functor such that:*

$H_r(A) = 0$ *if* $r > 0$.

$H_0(A) = A_G$.

$H_r(A) = 0$ *if A is projective in* $\mathrm{Mod}(G)$ *and* $r \geqq 1$.

The explicit determination of $H_0(A) = A_G$ comes from the fact that

$$X_1 \otimes_G A \to X_0 \otimes_G A \to \mathbf{Z} \otimes_G A \to 0$$

is exact, and that $\mathbf{Z} \otimes_G A$ is functorially isomorphic to A_G.

Given a short exact sequence $0 \to A' \to A \to A'' \to 0$, the long homology exact sequence starts

$$\cdots \to H_1(A'') \to A'_G \to A_G \to A''_G \to 0.$$

The previous two theorems fit a standard pattern of the derived functor. In some instances, we have to go back to the way these functors are constructed by means of complexes, say as in *Algebra*, Chapter XX, Theorem 2.1. We summarize this construction as follows for abelian categories.

Theorem 3.3. *Let $\mathfrak{A}, \mathfrak{B}$ be abelian categories. Let*

$$Y : \mathfrak{A} \to C(\mathfrak{B})$$

be an exact functor to the category of complexes in $\mathfrak{B}$. Then there exists a cohomological functor H on $\mathfrak{A}$ with values in $\mathfrak{B}$, such that $H^r(A) =$ homology of the complex $Y(A)$ in dimension r. Given a short exact sequence in $\mathfrak{A}$:

$$0 \to A' \xrightarrow{u} A \xrightarrow{v} A'' \to 0$$

and therefore the exact sequence

$$0 \to Y(A') \to Y(A) \to Y(A'') \to 0,$$

the coboundary is given by the usual formula $Y(u)^{-1} d Y(v)^{-1}$.

For the applications, readers may take $\mathfrak{B}$ to be the category of abelian groups, and $\mathfrak{A}$ is $\mathrm{Mod}(G)$ most of the time.

Corollary 3.4. *Let $\mathfrak{A}_1, \mathfrak{A}$ be abelian categories and F a bifunctor on $\mathfrak{A}_1 \times \mathfrak{A}$ with values in $\mathfrak{B}$, contravariant (resp. covariant) in the first variable and covariant in the second. Let X be a complex in $C(\mathfrak{A}_1)$ such that the functor $A \mapsto F(X, A)$ on $\mathfrak{A}$ is exact. Then there exists a cohomological functor (resp. homological functor) H on $\mathfrak{A}$ with values in $\mathfrak{B}$, obtained as in Theorem 3.3, with $F(X, A) = Y(A)$.*

Next we deal with finite groups, for which we obtain a non-trivial cohomological functor in all dimensions, using constructions with complexes as in the above two theorems.

Finite groups. Suppose now that G is finite, so we have the trace homomorphism

$$\mathbf{S} = \mathbf{S}_G : A \to A$$

for every $A \in \mathrm{Mod}(G)$. We omit the index G for simplicity, so the kernel of the trace in A is denoted by $A_{\mathbf{S}}$. We also write I instead of I_G as long as G is the only group under consideration. It is clear that IA is contained in $A_{\mathbf{S}}$ and the association

$$A \mapsto A_{\mathbf{S}}/IA$$

is a functor from $\mathrm{Mod}(G)$ into Grab. We then have **Tate's theorem.**

Theorem 3.5. *Let G be a finite group. There is a cohomological functor $\mathbf{H}$ on $mod(G)$ with values in* Grab *such that:*

$\mathbf{H}^0$ *is the functor* $A \mapsto A^G/\mathbf{S}_G A$.

$\mathbf{H}^r(A) = 0$ *if A is injective and $r \geqq 1$.*

$\mathbf{H}^r(A) = 0$ *if A is projective and r is arbitrary.*

$\mathbf{H}$ *is erased by G-regular modules, and thus is erased by* $\mathbf{M}_G$.

Proof. Fix the projective resolution X of $\mathbf{Z}$ and apply the two bifunctors $\otimes_G$ and Hom_G to obtain a diagram:

$$\rightarrow X_1 \otimes_G A \rightarrow X_0 \otimes_G A \qquad \mathrm{Hom}_G(X_0, A) \rightarrow \mathrm{Hom}_G(X_1, A) \rightarrow$$
$$\downarrow \qquad\qquad \downarrow$$
$$\mathbf{Z} \otimes_G A \qquad \mathrm{Hom}_G(\mathbf{Z}, A)$$
$$\downarrow \qquad\qquad \downarrow$$
$$0 \qquad\qquad 0$$

We have $A_G = \mathbf{Z} \otimes_G A$ and $\mathrm{Hom}_G(\mathbf{Z}, A) = A^G$. The right side with Hom_G comes from Theorem 3.1 and the left side comes from Theorem 3.2. We shall splice these two sides together. The trace maps $A_G \rightarrow A^G$ and yields a morphism of the functor $A \mapsto A_G$ to the functor $A \mapsto A^G$. Hence there exists a unique homomorphism δ which makes the following diagram commutative.

$$\rightarrow X_1 \otimes_G A \rightarrow X_0 \otimes_G^A \overset{\delta}{\rightarrow} \mathrm{Hom}_G(X_0, A) \rightarrow \mathrm{Hom}_G(X_1, A) \rightarrow$$
$$\downarrow \qquad\qquad\qquad \uparrow$$
$$\downarrow \qquad\qquad\qquad \uparrow$$
$$A_G \underset{\mathbf{S}_G}{\rightarrow} A^G$$
$$0 \qquad\qquad 0$$

The upper horizontal line is then a complex. Each $X_r \otimes_G A$ may be considered as a functor in A, and similarly for $\mathrm{Hom}_G(X_r, A)$. These functors are exact since X_r is projective. Furthermore, δ is a morphism of the functor $X_0 \otimes_G$ into the functor $\mathrm{Hom}_G(X_0, \cdot)$. We let

$$Y_r(A) = \begin{cases} \mathrm{Hom}_G(X_r, A) & \text{for } r \geqq 0 \\ X_{-r-1} \otimes_G A & \text{for } r < 0. \end{cases}$$

Then $Y(A)$ is a complex, and $A \mapsto Y(A)$ is exact, meaning that if $0 \rightarrow A' \rightarrow A \rightarrow A'' \rightarrow 0$ is a short exact sequence, then

$$0 \rightarrow Y(A') \rightarrow Y(A) \rightarrow Y(A'') \rightarrow 0$$

is exact. We are thus in the standard situation of constructing a homology functor, say as in *Algebra*, Chapter XX, Theorem 2.1, whereby $\mathbf{H}^r(A)$ is the homology in dimension r of the complex $Y(A)$. In dimensions 0 and -1, we find the functors of Theorem 3.1 and 3.2, thus proving all but the last statement of the theorem, concerning the erasability.

To show that G-regular modules erase the cohomology, we do it first in dimension $r > 0$. There exists a homotopy (Cf. *Algebra*, Chapter XX, §5), i.e. a family of $\mathbf{Z}$-morphisms

$$D_r : X_r \longrightarrow X_{r+1}$$

such that

$$\mathrm{id}_r = \mathrm{id}_{X_r} = \partial_{r+1} D_r + D_{r-1} \partial_r.$$

(Cf. the Remark at the end of §5, loc. cit.) Let $f : X_r \to A$ be a cocycle. By definition, $f\partial_{r+1} = 0$ and hence

$$f = f \circ \mathrm{id}_r = f D_{r-1} \partial_r.$$

On the other hand, by hypothesis there exists a $\mathbf{Z}$-morphism $v : A \to A$ such that $1_A = \mathbf{S}_G(v)$. Thus we find

$$f = 1_A f = \mathbf{S}_G(v)f = \mathbf{S}_G(vf) = \mathbf{S}_G(vfD_{r-1}\partial_r)$$
$$= \mathbf{S}_G(vfD_{r-1})\partial_r,$$

which shows that f is a coboundary, in other words, the cohomology group is trivial.

For $r = 0$ we obviously have $\mathbf{H}^0(A) = 0$ if A is G-regular. For $r = -1$, the reader will check it directly. For $r < -1$, one repeats the above argument for $r \geq 1$ with the tensor product, essentially dualizing the argument (reversing the arrows). This concludes the proof of Theorem 3.3.

Alternatively, the splicing used to prove Theorem 3.3 could also be done as follows, using a complete resolution of $\mathbf{Z}$.

Let $X = (X_r)_{r \geq 0}$ be a G-free resolution of $\mathbf{Z}$, with X_r finitely generated for all r, acyclic, with augmentation ε. Define

$$X_{-r-1} = \mathrm{Hom}(X_r, \mathbf{Z}) \text{ for } r \geq 0.$$

Thus we have defined G-modules X_s for negative dimensions s. One sees immediately that these modules are G-free. If (e_i) is a basis of X_r over $Z[G]$ for $r \geq 0$, then we have the dual basis $(e_i^\vee)$ for the dual module. By duality, we thus obtain G-free modules in negative dimensions. We may splice these two complexes around 0. For simplicity, say X_0 has dimension 1 over $Z[G]$ (which is the case in the complexes we select for the applications). Thus let

$$X_0 = Z[G] \cdot \xi, \text{ with } \varepsilon(\xi) = 1.$$

Let $\xi^\vee$ be the dual basis of $X_{-1} = \mathrm{Hom}(X_0, Z)$. We define ∂_0 by

$$\partial_0 \xi = S_G(\xi^\vee) = \sum_{\sigma \in G} \sigma \xi^\vee.$$

We can illustrate the relevant maps by the diagram:

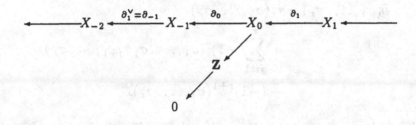

The boundaries ∂_{-r-1} for $r \geq 0$ are defined by duality. One verifies easily that the complex we have just obtained is acyclic (for instance by using a homotopy in positive dimension, and by dualising). We can then consider $\mathrm{Hom}_G(X, A)$ for A variable in $\mathrm{Mod}(G)$, and we obtain an exact functor

$$A \mapsto \mathrm{Hom}_G(X, A)$$

from $\mathrm{Mod}(G)$ into the category of complexes of abelian groups. In dimension 0, the homology of this complex is obviously

$$H^0(A) = A^G / S_G A,$$

and the uniqueness theorem applies.

The cohomological functor of Theorem 3.5 for finite groups will be called the **special** cohomology, to distinguish it from the cohomology defined for arbitrary groups, differing in dimension 0 and the negative dimensions. We write it as $\mathbf{H}_G$ if we need to specify G in the notation, especially when we shall deal with different groups and subgroups. It is uniquely determined, up to a unique isomorphism. When G is fixed throughout a discussion, we continue to denote it by $\mathbf{H}$. Thus depending on the context, we may write

$$\mathbf{H}(A) = \mathbf{H}_G(A) = \mathbf{H}(G, A) \text{ for } A \in \mathrm{Mod}(G).$$

The standard complex

The complex which we now describe allows for explicit computations of the cohomology groups. Let X_r be the free $\mathbf{Z}[G]$-module having for basis r-tuples $(\sigma_1, \ldots, \sigma_r)$ of elements of G. For $r = 0$ we take X_0 to be free over $\mathbf{Z}[G]$, of dimension 1, with basis element denoted by $(\cdot)$. We define the boundary maps by

$$\partial(\sigma_1, \ldots, \sigma_r) = \sigma_1(\sigma_2, \ldots, \sigma_r)$$
$$+ \sum_{j=1}^{r} (-1)^j (\sigma_1, \ldots, \sigma_j \sigma_{j+1}, \ldots, \sigma_r)$$
$$+ (-1)^{r+1} (\sigma_1, \ldots, \sigma_r).$$

We leave it to the reader to verify that $dd = 0$, i.e. we have a complex called the **standard complex**.

The above complex is the non-homogeneous form of another standard complex having nothing to do with groups. Indeed, let S be a set. For $r = 0, 1, 2, \ldots$ let E_r be the free $\mathbf{Z}$-module generated by $(r+1)$-tuples $(x_0, \ldots, x_r)$ with $x_0, \ldots, x_r \in S$. Thus such $(r+1)$-tuples form a basis of E_r over $\mathbf{Z}$. There is a unique homomorphism

$$d_{r+1} : E_{r+1} \to E_r$$

such that

$$d_{r+1}(x_0, \ldots, x_r) = \sum_{j=0}^{r+1} (-1)^j (x_0, \ldots, \hat{x}_j, \ldots, x_{r+1}),$$

where the symbol $\hat{x}_j$ means that this term is to be omitted. For $r = 0$ we define $\varepsilon : E_0 \to \mathbf{Z}$ to be the unique homomorphism such that $\varepsilon(x_0) = 1$. The map ε is also called the **augmentation**. Then we obtain a resolution of $\mathbf{Z}$ by the complex

$$\to E_{r+1} \to E_r \to \cdots \to E_0 \to \mathbf{Z} \to 0.$$

The above standard complex for an arbitrary set is called the **homogeneous standard complex**. It is exact, as one sees by using a homotopy as follows. Let $z \in S$ and define

$$h : E_r \to E_{r+1} \qquad \text{by} \qquad h(x_0, \ldots, x_r) = (z, x_0, \ldots, x_r).$$

Then it is routinely verified that

$$dh + hd = \text{id} \qquad \text{and} \qquad dd = 0.$$

Exactness follows at once.

Suppose now that the set S is the group G. Then we may define an action of G on the homogeneous complex E by letting

$$\sigma(\sigma_0, \ldots, \sigma_r) = (\sigma\sigma_0, \ldots, \sigma\sigma_r).$$

It is then routinely verified that each E_r is $\mathbf{Z}[G]$-free. We take $z = e$. Thus the homogeneous complex gives a $\mathbf{Z}[G]$-free resolution of $\mathbf{Z}$.

In addition, we have a $\mathbf{Z}[G]$-isomorphism $X \xrightarrow{\approx} E$ between the non-homogeneous and the homogeneous complex uniquely determined by the value on basis elements such that

$$(\sigma_1, \ldots, \sigma_r) \mapsto (e, \sigma_1, \sigma_1\sigma_2, \ldots, \sigma_1\sigma_2 \ldots \sigma_r).$$

The reader will immediately verify that the boundary operator ∂ given for X corresponds to the boundary operator as given on E under this isomorphism.

If G is finite, then each X_r is finitely generated. We may then proceed as was done in general to define the standard modules X_s in negative dimensions. The dual basis of $\{(\sigma_1, \ldots, \sigma_r)\}$ will be denoted by $\{[\sigma_1, \ldots, \sigma_r]\}$ for $r \geq 1$. The dual basis of $(\cdot)$ in dimension 0 will be denoted by $[\cdot]$. For finite groups, we thus obtain:

Theorem 3.6. *Let G be a finite group. Let $X = \{X_r\}(r \in \mathbf{Z})$ be the standard complex. Then X is $\mathbf{Z}[G]$-free, acyclic, and such that the association*

$$A \mapsto \mathrm{Hom}_G(X, A)$$

is an exact functor of $\mathrm{Mod}(G)$ into the category of complexes of abelian groups. The corresponding cohomological functor $\mathbf{H}$ is such that $\mathbf{H}^0(A) = A^G/S_G A$.

Examples. In the standard complex, the group of 1-cocycles consists of maps $f : G \to A$ such that

$$f(\sigma) + \sigma f(\tau) = f(\sigma\tau) \text{ for all } \sigma, \tau \in G.$$

The 1-coboundaries consist of maps f of the form $f(\sigma) = \sigma a - a$ for some $a \in A$. Observe that if G has trivial action on A, then by the above formulas,

$$H^1(A) = \mathrm{Hom}(G, A).$$

In particular, $H^1(\mathbf{Q}/\mathbf{Z}) = \hat{G}$ is the character group of G.

The 2-cocycles have also been known as **factor sets**, and are maps $f(\sigma, \tau)$ of two variables in G satisfying

$$f(\sigma, \tau) + f(\sigma\tau, \rho) = \sigma f(\tau, \rho) + f(\sigma, \tau\rho).$$

In Theorem 3.5, we showed that for finite groups, H is erased by M_G. The analogous statement in Theorem 3.1 has been left open. We can now settle it by using the standard complex.

Theorem 3.7. *Let G be any group. Let $B \in \mathrm{Mod}(\mathbf{Z})$. Then for all subgroups G' of G we have*

$$H^r(G', M_G(B)) = 0 \text{ for } r > 0.$$

Proof. By Proposition 2.6 it suffices to prove the theorem when $G' = G$. Define a map h on the chains of the standard complex by

$$h : C^r(G, M_G(B)) \to C^{r-1}(G, M_G(B))$$

by letting
$$(hf)_{\sigma_2,\ldots,\sigma_r}(\sigma) = f_{\sigma,\sigma_2,\ldots,\sigma_r}(e).$$

One verifies at once that
$$f = hdf + dhf,$$

whence the theorem follows. (Cf. *Algebra*, Chapter XX, §5.)

§4. Explicit computations

In this section we compute some low dimensional cohomology groups with some special coefficients.

We recall the exact sequences
$$0 \to I_G \to \mathbf{Z}[G] \to \mathbf{Z} \to 0$$

$$0 \to \mathbf{Z} \to \mathbf{Q} \to \mathbf{Q}/\mathbf{Z} \to 0.$$

We suppose G finite of order n. We write $I = I_G$ and $H = H_G$ for simplicity. We find:

$$
\begin{aligned}
\mathbf{H}^{-3}(\mathbf{Q}/\mathbf{Z}) &\approx \mathbf{H}^{-2}(\mathbf{Z}) &\approx \mathbf{H}^{-1}(I) &= I/I^2 \\
\mathbf{H}^{-2}(\mathbf{Q}/\mathbf{Z}) &\approx \mathbf{H}^{-1}(\mathbf{Z}) &\approx \mathbf{H}^{0}(I) &= 0 \\
\mathbf{H}^{-1}(\mathbf{Q}/\mathbf{Z}) &\approx \mathbf{H}^{0}(\mathbf{Z}) &\approx \mathbf{H}^{1}(I) &= \mathbf{Z}/n\mathbf{Z} \\
\mathbf{H}^{0}(\mathbf{Q}/\mathbf{Z}) &\approx \mathbf{H}^{1}(\mathbf{Z}) &\approx \mathbf{H}^{2}(I) &= 0 \\
\mathbf{H}^{1}(\mathbf{Q}/\mathbf{Z}) &\approx \mathbf{H}^{2}(\mathbf{Z}) &\approx \mathbf{H}^{3}(I) &= \hat{G}.
\end{aligned}
$$

The proof of these formulas arises as follows. Each middle term in the above exact sequences annuls the cohomological functor because $\mathbf{Z}[G]$ is G-regular in the first case, and $\mathbf{Q}$ is uniquely divisible in the second case. The stated isomorphisms other than those furthest to the right are then those induced by the coboundary in the cohomology exact sequence.

As for the values furthest to the right, they are proved as follows.

For the first one with I/I^2, we note that every element of I has trace 0, and hence $\mathbf{H}^{-1}(I) = I/I^2$ directly from its definition as in Theorem 2.7.

For the second line, we immediately have $\mathbf{H}^{-1}(\mathbf{Z}) = 0$ since an element of $\mathbf{Z}$ with trace 0 can only be equal to 0.

For the third line, we have obviously $\mathbf{H}^0(\mathbf{Z}) = \mathbf{Z}/n\mathbf{Z}$ from the definition.

For the fourth line, $\mathbf{H}^1(\mathbf{Z}) = 0$ because from the standard complex, this group consists of homomorphism from G into $\mathbf{Z}$, and G is finite.

For the fifth line, we find $\hat{G}$ because $\mathbf{H}^1(\mathbf{Q}/\mathbf{Z})$ is the dual group $\mathrm{Hom}(G, \mathbf{Q}/\mathbf{Z})$.

Remark. Let U be a subgroup of G. Then

$$\mathbf{H}^r(U, \mathbf{Z}[G]) = \mathbf{H}^r(U, \mathbf{Q}) = 0 \text{ for } r \in \mathbf{Z}.$$

Hence the above table applies also to a subgroup U, if we replace $\mathbf{H}_G$ by $\mathbf{H}_U$ and replace $n_G = \#(G)$ by $n_U = \#(U)$ in the third line, as well as G by U in the last line.

As far as I/I^2 is concerned, we have another characterization.

Proposition 4.1. *Let G be a group (possible infinite). Let G^c be its commutator group. Let I_G be the ideal of $\mathbf{Z}[G]$ generated by all elements of the form $\sigma - e$. Then there is a functorial isomorphism (covariant on the category of groups)*

$$G/G^c \approx I_G/I_G^2, \text{ given by } \sigma G^c \mapsto (\sigma - e) + I_G^2.$$

Proof. We can define a map $G \to I/I^2$ by $\sigma \mapsto (\sigma - e) + I^2$. One verifies at once that this map is a homomorphism. Since I/I^2 is commutative, G^c is contained in the kernel of the homomorphism, whence we obtain a homomorphism $G/G^c \to I/I^2$. Conversely, I is $\mathbf{Z}$-free, and the elements $(\sigma - e)$ with $\sigma \in G$, $\sigma \neq e$ form a basis over $\mathbf{Z}$. Hence there exists a homomorphism $I \mapsto G/G^c$ defined by the formula $(\sigma - e) \mapsto \sigma G^c, \sigma \neq e$. In addition, this homomorphism is trivial on I^2, as one verifies at once. Thus we obtain a homomorphism of I/I^2 into G/G^c, which is visibly inverse of the previous homomorphism of G/G^c into I/I^2. This proves the proposition.

In particular, from the first line of the table, we find the isomorphism

$$\mathbf{H}^{-2}(\mathbf{Z}) \approx G/G^c,$$

obtained from the coboundary and the isomorphism of Proposition 4.1. This isomorphism is important in class field theory.

We end our explicit computations with one more result on H^1.

Proposition 4.2. *Let G be a group, $A \in \mathrm{Mod}(G)$, and let $\alpha \in H^1(G, A)$. Let $\{a(\sigma)\}$ be a 1-cocycle representing α. There exists a G-morphism*

$$f : I_G \to A$$

such that $f(\sigma - e) = a(\sigma)$, i.e. one has $f \in (\mathrm{Hom}(I_G, A))^G$. Then the sequence

$$0 \to A = \mathrm{Hom}(\mathbf{Z}, A) \to \mathrm{Hom}(\mathbf{Z}[G], A) \to \mathrm{Hom}(I_G, A) \to 0$$

is exact, and taking the coboundary with respect to this short exact sequence, one has

$$\delta(\varkappa_G f) = -\alpha.$$

Proof. Since the elements $(\sigma - e)$ form a basis of I_G over $\mathbf{Z}$, one can define a $\mathbf{Z}$-morphism f satisfying $f(\sigma - e) = a(\sigma)$ for $\sigma \neq e$. The formula is even valid for $\sigma = e$, because putting $\sigma = \tau = e$ in the formula for the coboundary

$$a(\sigma\tau) = a(\sigma) + \sigma a(\tau),$$

we find $a(e) = 0$. We claim that f is a G-morphism. Indeed, for $\sigma, \tau \in G$ we find:

$$\begin{aligned}
f(\sigma(\tau - e)) = f(\sigma\tau - \sigma) &= f((\sigma\tau - e) - (\sigma - e)) \\
&= f(\sigma\tau - e) - f(\sigma - e) \\
&= a(\sigma\tau) - a(\sigma) \\
&= \sigma a(\tau) \\
&= \sigma f(\tau - e).
\end{aligned}$$

To compute $\varkappa_G f$ we first have to find a standard cochain of $\mathrm{Hom}(\mathbf{Z}[G], A)$ in dimension 0, mapping on f, that is an element $f' \in \mathrm{Hom}(\mathbf{Z}[G], A)$ whose restriction to I_G is f. Since

$$\mathbf{Z}[G] = I_G + \mathbf{Z}e$$

is a direct sum, we can define f' by prescribing that $f'(e) = 0$ and f' is equal to f on I_G. One then sees that $g_\sigma = \sigma f' - f'$ is a cocycle of dimension 1, in $\text{Hom}(\mathbf{Z}, A)$, representing $\varkappa_G f$ by definition. I claim that under the identification of $\text{Hom}(\mathbf{Z}, A)$ with A, the map g_σ corresponds to $-a(\sigma)$. In other words, we have to verify that $g_\sigma(e) = -a(\sigma)$. Here goes:

$$g_\sigma(e) = (\sigma f')(e) - f'(e)$$
$$= \sigma f'(\sigma^{-1})$$
$$= \sigma f(\sigma^{-1} - e)$$
$$= f(e - \sigma)$$
$$= -a(\sigma),$$

thus proving our assertion and concluding the proof of Proposition 4.2.

§5. Cyclic groups

Throughout this section we let G be a finite cyclic group, and we let σ be a generator of G

The main feature of the cohomology of such a cyclic group is that the cohomology is periodic of period 2, as we shall now prove.

We start with the δ-functor in two dimensions

$$\mathbf{H}_G^{-1} \text{ and } \mathbf{H}_G^0.$$

Recall that $\varkappa : A^G \to \mathbf{H}^0(G, A) = A^G/S_G A$ and

$$\varkappa : A_{S_G} \to \mathbf{H}^{-1}(G, A) = A_{S_G}/I_G A$$

are the canonical homomorphisms. We are going to define a cohomological functor directly from these maps. For $r \in \mathbf{Z}$ we let:

$$\mathbf{H}^r(G, A) = \begin{cases} \mathbf{H}^{-1}(G, A) & \text{if } r \text{ is odd} \\ \mathbf{H}^0(G, A) & \text{if } r \text{ is even.} \end{cases}$$

We then have to define the coboundary. Let

$$0 \to A' \xrightarrow{u} A \xrightarrow{v} A'' \to 0$$

be a short exact sequence in Mod(G). For each $r \in \mathbf{Z}$ we define $\varkappa_r$ and $\bar{\varkappa}_r$ in the natural way given the above definition, and for $a'' \in A''^G$ we pick any element $a \in A$ such that $va = a''$ and define

$$\delta_\sigma \varkappa_r(a'') = \bar{\varkappa}_r(\sigma a - a).$$

Similarly, in odd dimensions, for $a'' \in A_{S_G}$ we pick $a \in A$ such that $va = a''$ and we define

$$\delta_\sigma \bar{\varkappa}_r(a'') = \varkappa_r(\mathbf{S}_G a).$$

It is immediately verified that δ_σ is well-defined (depending on the choice of generator σ), that is independent of the choice of a such that $va = a''$. It is then also routinely and easily verified that the sequence $\{H^r\}$ ($r \in \mathbf{Z}$) with the coboundary δ_σ is a cohomological functor. Since it vanishes on G-regular modules, and is given as before in dimension < 0, it follows that it is isomorphic to the special functor defined previously, by the uniqueness theorem. Directly from this new definition, we now see that for all $r \in \mathbf{Z}$ and $A \in \text{Mod}(G)$ we have the periodicity

$$\mathbf{H}^{r+2}(G, A) = \mathbf{H}^r(G, A).$$

Of course, by truncating on the left we can define a similar functor in the ordinary case. We put $H^0(A) = A^G$ as before, and for $r \geq 1$ we let:

$$H^r(A) = \begin{cases} H^{-1}(G, A) & \text{if } r \text{ is odd} \\ H^0(G, A) & \text{if } r \text{ is even.} \end{cases}$$

We define the coboundary as before, so we find a cohomological functor which is periodic for $r \geq 0$, and 0 in negative dimensions. Again the uniqueness theorem shows that it coincides with the functor defined in the previous sections. The beginning of the cohomology sequence reads:

$$0 \to A'^G \to A^G \to A''^G \to H^{-1}(G, A)$$

and it continues as for the special functor.

The cohomology sequence for the special functor can be conveniently written as in the next theorem.

34

Theorem 5.1. *Let G be finite cyclic and let σ be a generator. Let*

$$0 \to A' \to A \to A'' \to 0$$

be a short exact sequence in $\mathrm{Mod}(G)$. *Then the following hexagon is exact:*

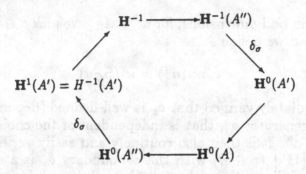

Suppose given an exact hexagon of finite abelian groups as shown:

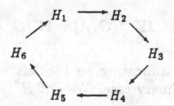

and let h_i be the order of H_i, that is $h_i = (H_i : 0)$. Let

$$f_i : H_i \to H_{i+1} \qquad \text{with} \qquad i \bmod 6$$

be the corresponding homomorphism in the diagram. Then

$$h_i = (H_i : f_{i-1}H_{i-1})(f_{i-1}H_{i-1} : 0) = m_i m_{i-1}.$$

Hence

$$1 = \frac{m_0 m_1 m_2 m_3 m_4 m_5}{m_1 m_2 m_3 m_4 m_5 m_6} = \frac{h_1 h_3 h_5}{h_2 h_4 h_6}.$$

We may apply this formula to the exact sequence of Theorem 5.1. Assume that each group $\mathbf{H}^i(A)$ is finite, and let $h_i(A) = $ order of $\mathbf{H}^i(G, A)$. Then

$$1 = \frac{h_1(A')h_1(A'')h_2(A)}{h_1(A)h_2(A')h_2(A'')}.$$

Now let $A \in \text{Mod}(G)$ be arbitrary. If $h_1(A)$ and $h_2(A)$ are finite, we define the **Herbrand quotient** $h_{2/1}(A)$ to be

$$h_{2/1}(A) = \frac{h_2(A)}{h_1(A)} = \frac{(A_{\sigma - e} : \mathbf{S}_G A)}{A_{\mathbf{S}_G} : (\sigma - e)A}.$$

If $h_1(A)$ or $h_2(A)$ is not finite, we say that the Herbrand quotient is **not defined**. This Herbrand quotient is in fact a Euler characteristic, cf. for instance *Algebra*, Chapter XX, §3. If A is a finite abelian group in $\text{Mod}(G)$, then the Herbrand quotient is defined. The main properties of the Herbrand quotient are contained in the next theorems.

Theorem 5.2. Herbrand's Lemma. *Let G be a finite cyclic group, and let*

$$0 \rightarrow A' \rightarrow A \rightarrow A'' \rightarrow 0$$

be a short exact sequence in $\text{Mod}(G)$. *If two out of three Herbrand quotients* $h_{2/1}(A'), h_{2/1}(A), h_{2/1}(A'')$ *are defined so is the third, and one has*

$$h_{2/1}(A) = h_{2/1}(A')h_{2/1}(A'').$$

Proof. This follows at once from the discussion preceding the theorem. It is also a special case of *Algebra*, Chapter XX, Theorem 3.3.

Theorem 5.3. *Let G be finite cyclic and suppose $A \in \text{Mod}(G)$ is finite. Then $h_{2/1}(A) = 0$.*

Proof. We have a lattice of subgroups:

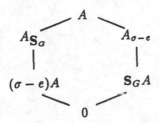

The factor groups $A/A_{\mathbf{S}_G}$ and $\mathbf{S}_G A$ are isomorphic, and so are $A/A_{\sigma - e}$ and $(\sigma - e)A$. Computing the order $(A : 0)$ going around both sides of the hexagon, we find that the factor groups of the two vertical sides have the same order, that is

$$(A_{\mathbf{S}_G} : (\sigma - e)A) = (A_{\sigma - e} : \mathbf{S}_G A).$$

That $h_{2/1}(A) = 1$ now follows from the definitions.

Finally we have a result concerning the Herrbrand quotient for trivial action.

Theorem 5.4. *Let G be a finite cyclic group of prime order p. Let $A \in \mathrm{Mod}(G)$. Let $t(A)$ be the Herbrand quotient relative to the trivial action of G on the abelian group A, so that*

$$t(A) = \frac{(A_p : 0)}{(A/pA : 0)}.$$

Suppose this quotient is defined. Then $t(A^G), t(A_G),$ and $h_{2/1}(A)$ are defined, and one has

$$h_{2/1}(A)^{p-1} = t(A^G)^p / t(A) = t(A_G)^p / t(A).$$

Proof. We leave the proof as an exercise (not completely trivial).

CHAPTER II
Relations with Subgroups

This chapter tabulates systematically a number of relations between the cohomology of a group and that of its subgroups and factor groups.

§1. Various morphisms

(a) **Changing the group** G. Let $\lambda : G' \to G$ be a group homomorphism. Then λ gives rise to an exact functor

$$\Phi_\lambda : \mathrm{Mod}(G) \to \mathrm{Mod}(G')$$

because every G-module can be considered as a G'-module if we define the action of an element $\sigma' \in G'$ on an element $a \in A$ by

$$\sigma' a = \lambda(\sigma')a.$$

We may therefore consider the cohomological functor $H_{G'} \circ \Phi_\lambda$ (or the special functor $\mathbf{H}_{G'} \circ \Phi_\lambda$ if G' is finite) on $\mathrm{Mod}(G)$.

In dimension 0, we have a morphism of functors

$$H_G^0 \to H_{G'}^0 \circ \Phi_\lambda$$

given by the inclusion $A^G \hookrightarrow A^{G'} = (\Phi_\lambda(A))^{G'}$. If in addition G and G' are finite, then we have a morphism of functors

$$\mathbf{H}_G^0 \to \mathbf{H}_{G'}^0 \circ \Phi_\lambda$$

given by the homomorphism $A^G/S_G A \to A^{G'}/S_{G'} A$, with G' acting on A in the manner prescribed above, via λ.

By the uniqueness theorem, there exists a unique morphism of cohomological functor (δ-morphism)

$$\lambda^* : H_G \to H_{G'} \circ \Phi_\lambda \quad \text{or} \quad \mathbf{H}_G \to \mathbf{H}_{G'} \circ \Phi_\lambda,$$

the second possibility arising when G and G' are finite. We shall now make this map λ^* explicit in various special cases.

Suppose λ is surjective. Then we call λ^* the **lifting** morphism, and we denote it by $\mathrm{lif}_{G'}^G$. In this case, G may be viewed as a factor group of G' and the lifting goes from the factor group to the group. On the other hand, when G' is a subgroup of G, then λ^* will be called the **restriction**, and will be studied in detail below.

Let $A \in \mathrm{Mod}(G)$ and $B \in \mathrm{Mod}(G')$. We may consider A as a G'-module as above (via the given λ). Let $v : A \to B$ be a G'-morphism. Then we say that the pair (λ, v) is a **morphism** of (G, A) to (G', B). One can define formally a category whose objects are pairs (G, A) for which the morphisms are precisely the pairs (λ, v). Every morphism (λ, v) induces a homomorphism

$$(\lambda, v)_* : H^r(G, A) \to H^r(G', B),$$

and similarly replacing H by the special $\mathbf{H}$ if G and G' are finite, by taking the composite

$$\mathbf{H}^r(G, A) \xrightarrow{\lambda^*} \mathbf{H}^r(G', A) \xrightarrow{\mathbf{H}_{G'}(v)} \mathbf{H}^r(G', B).$$

Of course, we should write more correctly $\mathbf{H}^r(G', \Phi_\lambda(A))$, but usually we delete the explicit reference to Φ_λ when the reference is clear from the context.

Proposition 1.1. *Let (λ, v) be a morphism of (G, A) to $G', B)$, and let (φ, w) be a morphism of (G', B) to (G'', C). Then $(\lambda\varphi, wv)$ is a morphism of (G, A) to (G'', C), and one has*

$$(\varphi\lambda, wv)_* = (\varphi, w)_*(\lambda, v)_*.$$

Proof. Since φ^* is a morphism of functors, the following diagram is commutative:

$$
\begin{array}{ccc}
H^r(G',\Phi_\lambda(A)) & \xrightarrow{\ H_{G'}(v)\ } & H^r(G',B) \\
{\scriptstyle\varphi^*}\downarrow & & \downarrow{\scriptstyle\varphi^*} \\
H^r(G'',\Phi_\varphi\Phi_\lambda(A)) & \xrightarrow[\ H_{G''}(v)\ ]{} & H^r(G'',\Phi_\varphi(B)).
\end{array}
$$

Consequently we find

$$
\begin{aligned}
(\lambda\varphi, wv)_* &= H_{G''}(wv)\circ(\lambda\varphi)^* \\
&= H_{G''}(w)\circ H_{G''}(v)\circ\varphi^*\circ\lambda^* \\
&= H_{G''}(w)\circ\varphi^* H_{G'}(v)\circ\lambda^* \\
&= (\varphi,w)_*(\lambda,v)_*,
\end{aligned}
$$

thus proving the proposition.

(b) Restriction. This is the case when λ is an injection, so that we may consider G' as a subgroup of G. We therefore have for $A \in \mathrm{Mod}(G)$:

$$
\mathrm{res}^G_{G'} : H^r(G,A) \to H^r(G',A),
$$

and similarly for the special functor $\mathbf{H}^r$ when G is finite. One verifies at once that for $r > 0$ the restriction homomorphism is obtained from the standard complex by restricting a cochain $\{f(\sigma_1,\ldots,\sigma_r)\}$ as a function of r-tuples of elements of G to r-tuples of elements in G', because the restriction is a morphism of cohomological functors to which we can apply the uniqueness theorem. In dimension 0, the restriction is induced by the inclusion mapping.

We have transitivity:

Proposition 1.2. *Let $G' \supset G''$ be subgroups of G. Then on H_G, or $\mathbf{H}_G$ if G is finite, we have*

$$
\mathrm{res}^{G'}_{G''}\circ\mathrm{res}^G_{G'} = \mathrm{res}^G_{G''}.
$$

Proof. Immediate from Proposition 1.1.

(c) **Inflation.** Let $\lambda : G \to G/G'$ be a surjective homomorphism. Let $A \in \text{Mod}(G)$. Then $A^{G'}$ is a G/G'-module for the obvious action induced by the action of G, trivial on G', and of course $A^{G'}$ is also a G-module for this operation. We have a morphism of inclusion

$$u : A^{G'} \to A$$

in $\text{Mod}(G)$, which induces a homomorphism

$$H^r_G(u) = u_r : H^r(G, A^{G'}) \to H^r(G, A) \text{ for } r \geqq 0.$$

We define inflation

$$\text{inf}^{G/G'}_G : H^r(G/G', A^{G'}) \to H^r(G, A)$$

to be the composite of the functorial morphism

$$H^r(G/G', A^{G'}) \to H^r(G, A^{G'})$$

followed by the induced homomorphism u_r for $r \geqq 0$. Note that inflation is NOT defined for the special cohomology functor when G is finite.

In dimension 0, the inflation therefore gives the identity map

$$(A^{G'})^{G/G'} \to A^G.$$

In dimension $r > 0$, it is induced by the cochain homomorphism in the standard complex, which to each cochain $\{f(\bar{\sigma}_1, \ldots, \bar{\sigma}_r)\}$ with $\bar{\sigma}_i \in G/G'$ associates the cochain $\{f(\sigma_1, \ldots, \sigma_r)\}$ whose values are constant on cosets of G'.

We have already observed that if G acts trivially on A, then $H^1(G, A)$ is simply $\text{Hom}(G, A)$. Therefore we obtain:

Proposition 1.3. *Let G' be a normal subgroup of G and suppose G acts trivially on A. Then the inflation*

$$\text{inf}^{G/G'}_G : H^1(G/G', A^{G'}) \to H^1(G, A)$$

induces the inflation of a homomorphism $\bar{\chi} : G/G' \to A$ to a homomorphism $\chi : G \to A$.

Let G' be a normal subgroup of G. We may consider the association

$$F_G : A \mapsto A^{G'}$$

as a functor, not exact, from $\mathrm{Mod}(G)$ to $\mathrm{Mod}(G/G')$. Inflation is then a morphism of functors (but not a cohomological morphism)

$$H_{G/G'} \circ F_{G'} \to H_G$$

on the category $\mathrm{Mod}(G)$. Even though we are not dealing with a cohomological morphism, we can still use the uniqueness theorem to prove certain commutativity formulas, by decomposing the inflation into two pieces.

As another special case of Proposition 1.1, we have:

Proposition 1.4. *Let G', N be subgroups of G with N normal in G, and N contained in G'. Then on $H^r(G/N, A^N)$ we have*

$$\inf_{G'}^{G'/N} \circ \operatorname{res}_{G'/N}^{G/N} = \operatorname{res}_{G'}^G \circ \inf_G^{G/N}.$$

We also have transitivity, also as a special case of Proposition 1.1.

Proposition 1.4. *Let $G \to G_1 \to G_2$ be surjective group homomorphisms. Then*

$$\inf_G^{G_1} \circ \inf_{G_1}^{G_2} = \inf_G^{G_2}.$$

(d) Conjugation. Let U be a subgroup of G. For $\sigma \in G$ we have the conjugate subgroup

$$U^\sigma = \sigma^{-1} U \sigma = U[\sigma] = [\sigma^{-1}]U.$$

The notation is such that $U^{(\sigma\tau)} = (U^\sigma)^\tau$. On $\mathrm{Mod}(G)$ we have two cohomological functors, H_U and H_{U^σ}. In dimension 0, we have an isomorphism of functors

$$A^U \to A^{U^\sigma} \quad \text{given by} \quad a \mapsto \sigma^{-1}a.$$

We may therefore extend this isomorphism uniquely to an isomorphism of H_U with H_{U^σ} which we denote by σ_* and which we call **conjugation.**

Similarly if U is finite, we have conjugation σ_* on the special functor $H_U \to H_{U^\sigma}$.

Proposition 1.5. *If $\sigma \in U$ then σ_* is the identity on H_U (resp. $\mathbf{H}_U$ if U is finite).*

Proof. The assertion is true in dimension 0, whence in all dimensions.

Let $f : A \to B$ be a U-morphism with $A, B \in \mathrm{Mod}(G)$. Then

$$f^\sigma = [\sigma^{-1}]f = f[\sigma] : A \to B$$

is a U^σ-morphism. The fact that σ_* is a morphism of functors shows that

$$H_{U^\sigma}(f^\sigma) \circ \sigma_* = \sigma_* \circ H_U(f)$$

as morphisms on $H(U, A)$ (and similarly for H replaced by $\mathbf{H}$ if U is finite).

If U is a normal subgroup of G, then σ_* is an automorphism of H_U (resp. $\mathbf{H}_U$ if U is finite). In other words, G acts on H_U (or $\mathbf{H}_U$). Since we have seen that σ_* is trivial if $\sigma \in U$ it follows that actually G/U acts on H_U (resp. $\mathbf{H}_U$).

Proposition 1.7. *Let $V \subset U$ be subgroups of G, and let $\sigma \in G$. Then*

$$\sigma_* \circ \mathrm{res}_V^U = \mathrm{res}_{V^\sigma}^{U^\sigma} \circ \sigma_*$$

on H_U (resp. $\mathbf{H}_U$ if U is finite).

Proposition 1.8. *Let $V \subset U$ be subgroups of G of finite index, and let $\sigma \in G$. Suppose V normal in U. Then*

$$\mathrm{inf}_{U^\sigma}^{U^\sigma/V^\sigma} \circ \sigma_* = \sigma_* \circ \mathrm{inf}_U^{U/V}$$

on $H(U/V, A^V)$, with $A \in \mathrm{Mod}(G)$.

Both the above propositions are special cases of Proposition 1.1.

(e) The transfer. Let U be a subgroup of G, of finite index. The trace gives a morphism of functors $H_U^0 \to U_G^0$ by the formula

$$S_G^U : A^U \to A^G,$$

and similarly in the special case when G is finite, $\mathbf{H}_U^0 \to \mathbf{H}_G^0$ by

$$S_G^U : A^U/S_U A \to A^G/S_G A.$$

The unique extension to the cohomological functors will be denoted by tr_G^U, and will be called the **transfer**. The following proposition is proved by verifying the asserted commutativity in dimension 0, and then applying the uniqueness theorem. In the case of inflation, we decompose this map in its two components.

Proposition 1.9. *Let $V \subset U \subset G$ be subgroups of finite index in G. Then on H_V (resp. $\mathbf{H}_V$) we have*

(1) $\mathrm{tr}_G^U \circ \mathrm{tr}_U^V = \mathrm{tr}_G^V$.

(2) $\sigma_* \circ \mathrm{tr}_U^V = \mathrm{tr}_U^V \circ \sigma_*$ *for $\sigma \in G$.*

(3) *If V is normal in G, then on $H^r(U/V, A^V)$ with $r \geqq 0$ we have*

$$\mathrm{inf}_G^{G/V} \circ \mathrm{tr}_{G/V}^{U/V} = \mathrm{tr}_G^U \circ \mathrm{inf}_U^{U/V}.$$

The next result is particularly important.

Proposition 1.10. *Let U be a subgroup of finite index in G. Then on H_G (resp. $\mathbf{H}_G$) we have*

$$\mathrm{tr}_G^U \circ \mathrm{res}_U^G = (G : U),$$

where $(G : U)$ on the right abbreviates $(G : U)_H$, i.e. multiplication by the index on the cohomology functor.

Proof Again the formula is immediate in dimension 0, since restriction is just inclusion, and so the trace simply multiplies elements by $(G : U)$. Then the proposition follows in general by applying the uniqueness theorem.

Corollary 1.11. *Suppose G finite of order n. Then for all $r \in \mathbf{Z}$ and $A \in \mathrm{Mod}(G)$ we have $n\mathbf{H}^r(G, A) = 0$.*

Proof. Take $U = \{e\}$ in the proposition and use the fact that $\mathbf{H}^r(e, A) = 0$.

Corollary 1.12. *Suppose G finite, and $A \in \mathrm{Mod}(G)$ finitely generated over $\mathbf{Z}$. Then $\mathbf{H}^r(G, A)$ is a finite group for all $r \in \mathbf{Z}$.*

Proof. First $\mathbf{H}^r(G, A)$ is finitely generated, because in the standard complex, the cochains are determined by their values on the finite number of generators of the complex in each idmension. Since $\mathbf{H}^r(G,A)$ is a torsion group by the preceding corollary, it follows that it is finite.

Corollary 1.13. *Suppose G finite and $A \in \mathrm{Mod}(G)$ is uniquely divisible by every integer $m \in \mathbf{Z}, m \neq 0$. Then $\mathbf{H}^r(G,A) = 0$ for all $r \in \mathbf{Z}$.*

Proposition 1.14. *Let $U \subset G$ be a subgroup of finite index. Let $A, B \in \mathrm{Mod}(G)$ and let $f : A \to B$ be a U-morphism. Then*

$$H_G(\mathbf{S}_G^U(f)) = \mathrm{tr}_G^U \circ H_U(f) \circ \mathrm{res}_U^G,$$

and similarly with $\mathbf{H}$ instead of H when G is finite.

Proof. We use the fact that the assertion is immediate in dimension 0, together with the technique of dimension shifting. We also use Chapter I, Lemma 2.4, that we can take a G-morphism in and out of a trace, so we find a commutative diagram

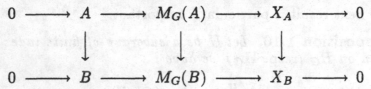

the three vertical maps being $\mathbf{S}_G^U(f)$, $\mathbf{S}_G^U(M(f))$ and $\mathbf{S}_G^U(X(f))$ respectively. In the hypothesis of the proposition, we replace f by $X(f) : X_A \to X_B$, and we suppose the proposition proved for $X(f)$. We then have two squares which form the faces of a cube as shown:

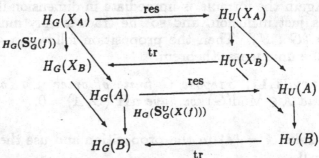

The maps going forward are the coboundary homomorphisms, and are surjective since M_G erases cohomology. Thus the diagram allows an induction on the dimension to conclude the proof. In the case of the special functor H, we use the dual diagram going to the left for the induction.

Corollary 1.15. *Suppose G finite and $A, B \in \mathrm{Mod}(G)$. Let $f : A \to B$ be a $\mathbf{Z}$-morphism. Then $\mathbf{S}_G(f) : A \to B$ induces 0 on all the cohomology groups.*

Proof. We can take $U = \{e\}$ in the preceding proposition.

Explicit formulas

We shall use systematically the following notation. We let $\{c\}$ be the set of right cosets of a subgroup U of G (not necessarily finite). We choose a set of coset representatives denoted by $\bar{c}$. If $\sigma \in G$, we denote by $\bar{\sigma}$ the representative of $U\sigma$. We may then write

$$G = \bigcup_c U\bar{c} = \bigcup_c \bar{c}^{-1}U$$

since $\{\bar{c}^{-1}\}$ is a system of representatives for the left cosets of U in G. By definition, we have $U\bar{c} = Uc$, whence for all $\sigma \in G$, we have

$$\bar{c}\sigma\overline{\bar{c}\sigma}^{-1} \in U.$$

We now give the explicit formula for the transfer on standard cochains. It is induced by the cochain map $f \mapsto \mathrm{tr}_G^U(f)$ given by

$$\mathrm{tr}_G^U(f)(\sigma_0, \ldots, \sigma_r) = \sum_c \bar{c}^{-1} f(\bar{c}\sigma_0\overline{\bar{c}\sigma_0}^{-1}, \ldots, \bar{c}\sigma_r\overline{\bar{c}\sigma_r}^{-1}).$$

For a non-homogeneous cochain, we have the formula

$$\mathrm{tr}_G^U(f)(\sigma_1, \ldots, \sigma_r) =$$
$$\sum_c \bar{c}^{-1} f(\bar{c}\sigma_1\overline{\bar{c}\sigma_1}^{-1}, \ldots, \overline{\bar{c}\sigma_1\sigma_2}\,\overline{\bar{c}\sigma_1\sigma_2}^{-1}, \ldots, \overline{\bar{c}\sigma_1\cdots\sigma_{r-1}}\sigma_r\overline{\bar{c}\sigma_1\cdots\sigma_r}^{-1}).$$

(f) Translation. Let G be a group, U a subgroup and N a normal subgroup of G. Let $A \in \mathrm{Mod}(G)$. Then we have a lattice of submodules of A:

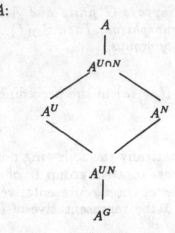

We have $UN/N \approx U/(U \cap N)$, and U acts on A^N since G acts on A^N. Furthermore $U \cap N$ leaves A^N fixed, and so we have a homomorphism called **translation**

$$\text{tsl}_* : H^r(UN/N, A^N) \to H^r(U/(U \cap N), A^{U \cap N})$$

for $r \geq 0$. The isomorphism $UN/N \approx U/(U \cap N)$ is compatible with the inclusion of A^N in $A^{U \cap N}$. Similarly, if G is finite, we get the translation for the special cohomology $\mathbf{H}$ instead of H, with $r \geq 0$.

Taking G arbitrary and $r \geq 0$, we have a commutative diagram:

$$H^r(G/N, A^N) \xrightarrow{\ \text{res}\ } H^r(UN/N, A^N) \xrightarrow{\ \text{inf}\ } H^r(UN, A)$$

$$\text{inf} \downarrow \qquad\qquad \downarrow \text{tsl} \qquad\qquad \downarrow \text{res}$$

$$H^r(U/(U \cap N), A^{U \cap N})$$

$$\downarrow \text{inf}$$

$$H^r(G, A) \xrightarrow[\ \text{res}\ ]{} H^r(U, A) \xrightarrow{\qquad} H^r(U, A)$$

The composition, which one can achieve in three ways,

$$\text{tsl}_* : H^r(G/N, A^N) \to H^r(U, A)$$

will also be called **translation**, and is denoted tsl_*.

In dimension -1, we have the following explicit determination of cohomology.

Proposition 1.16. *Let G be a finite group and U a subgroup. Let $A \in \mathrm{Mod}(G)$.*

(1) *For $a \in A_{S_G}$ we have $S_G^U(a) \in A_{S_U}$ and*

$$\mathrm{res}_U^G \varkappa_G(a) = \varkappa(S_G^U(a)).$$

(2) *For $a \in A_{S_U}$ we have $a \in A_{S_G}$ and*

$$\mathrm{tr}_G^U \varkappa_U(a) = \varkappa_G(a).$$

(3) *Let $a \in A_{S_U}$ and $\sigma \in G$. Then $\sigma^{-1}a \in A_{S_{U^\sigma}}$ and*

$$\sigma_* \varkappa_U(a) = \varkappa_{U^\sigma}(\sigma^{-1}a).$$

Proof. In each case, one verifies explicitly that the morphism of the functor $\mathbf{H}^{-1}$ given by the expression on the right of the formulas is a δ-morphism, for the pair of functors $(\mathbf{H}^{-1}, \mathbf{H}^0)$. We can then apply the uniqueness theorem. The verification is done routinely, and is left to the reader, who will use the explicit determination of δ in Chapter I, Theorem 2.7.

Roughly speaking, Proposition 1.16 asserts that the restriction and transfer correspond respectively to the trace and the inclusion (so the order is reversed with respect to dimension 0). Conjugation just consists in applying σ^{-1} to a cochain representing a cohomology class.

In Chapter I, §4 we gave an explicit determination of $\mathbf{H}^{-2}(G, \mathbf{Z})$. We shall now indicate how the transfer, restriction and conjugation behave with respect to this determination.

We recall the isomorphisms:

$$\mathbf{H}^{-2}(G, \mathbf{Z}) \xrightarrow[\approx]{\delta} \mathbf{H}^{-1}(G, I_G) = I_G/I_G^2 \xrightarrow{\approx} G/G^c.$$

If $\tau \in G$, we denote by ζ_τ the element of $\mathbf{H}^{-2}(G, \mathbf{Z})$ which corresponds to τG^c under the above isomorphism.

Directly in terms of groups, we have some natural homomorphisms as follows. If $\lambda : G \to \bar{G}$ is a homomorphism, then we have an induced homomorphism

$$\lambda^c : G/G^{c\cdot} \to \bar{G}/\bar{G}^c.$$

In particular, if U is a subgroup of G, we have the canonical homomorphism

$$\text{inc}_* : U/U^c \to G/G^c$$

induced by the inclusion.

If U is of finite index in G then we have the transfer from group theory

$$\text{Tr}_U^G = \text{Tr}_{U/U^c}^{G/G^c} : G/G^c \to U/U^c$$

defined by the product

$$\text{Tr}_U^G(\sigma G^c) = \prod_c (\bar{c}\sigma\bar{c\sigma}^{-1} U^c).$$

Cf. for instance Artin-Tate, Chapter XIII, §4.

We shall now see that the transfer and restriction on $\mathbf{H}^{-2}$ correspond to the inclusion and transfer on the groups (so the order is reversed).

Theorem 1.17. *Let G be a finite group and U a subgroup. Then:*

1. *The transfer* $\text{tr}_G^U : \mathbf{H}^{-2}(U, \mathbf{Z}) \to \mathbf{H}^{-2}(G, \mathbf{Z})$ *corresponds to the natural map* $U/U^c \to G/G^c$ *induced by the inclusion of U in G. Thus we may write*

$$\text{tr}(\zeta_\tau) = \zeta_\tau.$$

2. *The restriction* $\text{res}_U^G : \mathbf{H}^{-2}(G, \mathbf{Z}) \to \mathbf{H}^{-2}(U, \mathbf{Z})$ *corresponds to the transfer of group theory. Thus we may write*

$$\text{res}_U^G(\zeta_\sigma) = \zeta_{\text{tr}(\sigma)}.$$

3. *Conjugation* $\sigma_* : \mathbf{H}^{-2}(U, \mathbf{Z}) \to \mathbf{H}^{-2}(U^\sigma, \mathbf{Z})$ *corresponds to the map of U/U^c into $U^\sigma/(U^\sigma)^c$ induced by conjugation with $\sigma \in G$, so we may write*

$$\sigma_*(\zeta_\tau) = \zeta_{\sigma\tau\sigma^{-1}}.$$

Proof. Since $\mathbf{Z}[U]$ is naturally contained in $\mathbf{Z}[G]$ we obtain a commutative diagram

$$
\begin{array}{ccccccccc}
0 & \longrightarrow & I_U & \longrightarrow & \mathbf{Z}[U] & \longrightarrow & \mathbf{Z} & \longrightarrow & 0 \\
 & & \downarrow {\scriptstyle \text{inc}} & & \downarrow {\scriptstyle \text{inc}} & & \downarrow {\scriptstyle \text{id}} & & \\
0 & \longrightarrow & I_G & \longrightarrow & \mathbf{Z}[G] & \longrightarrow & \mathbf{Z} & \longrightarrow & 0
\end{array}
$$

the vertical maps being inclusions, and the map on $\mathbf{Z}$ being the identity. The horizontal sequences are exact. Consequently we obtain a commutative diagram

$$
\begin{array}{ccc}
H^{-2}(U,\mathbf{Z}) & \xrightarrow{\;\delta\;} & H^{-1}(U,I_U) \approx I_U/I_U^2 \\
\downarrow {\scriptstyle \text{id}} & & \downarrow {\scriptstyle \text{inc}_*} \\
H^{-2}(U,\mathbf{Z}) & \xrightarrow{\;\delta\;} & H^{-1}(U,I_G) \approx (I_G)_{\mathbf{s}_U}/I_U I_G.
\end{array}
$$

The coboundaries are isomorphisms, and hence inc_* is also an isomorphism. Thus we may write, as we have done, $(I_G)_{\mathbf{s}_U}/I_U I_G$ instead of I_U/I_U^2. In dimension -1 we may then use the explicit determination of H^{-2} from the preceding proposition, which we do case by case.

Let $\tau \in U$. Then $\tau - e$ is in $(I_G)_{\mathbf{s}_U}$, and we have

$$
\text{tr}_G^U \ae_U(\tau - e) = \ae_G(\tau - e),
$$

which proves the first formula.

Next let $\sigma \in G$. We have

$$
\text{res}_U^G \ae_U(\sigma - e) = \ae_U \left(\sum_c \bar{c}(\sigma - e) \right)
$$

where c ranges over the right cosets of U, and $\bar{c}$ is a representative of c. Furthermore

$$
\sum_c \bar{c}(\sigma - e) = \sum_c (\bar{c}\sigma - \overline{c\sigma})
$$

because $c \mapsto c\sigma$ permutes the cosets. Since $\bar{c}\sigma\overline{c\sigma}^{-1}$ is in U, we may rewrite this equality in the form

$$\sum_c (\bar{c}\sigma\overline{c\sigma}^{-1} - e)(\overline{c\sigma} - e) + \sum_c (\bar{c}\sigma\overline{c\sigma}^{-1} - e)$$

$$\equiv \sum_c (\bar{c}\sigma\overline{c\sigma}^{-1} - e) \mod I_U I_G.$$

If we apply ж to both sides, one sees that the second formula is proved, taking into account the formula for the transfer in group theory, which gives

$$\mathrm{Tr}_U^G(\sigma G^c) = \prod_c (\bar{c}\sigma\overline{c\sigma}^{-1} U^c).$$

As to the third formula, it is proved similarly, using the equalities

$$\sigma^{-1}(\sigma\tau\sigma^{-1} - e) = \tau\sigma^{-1} - \sigma^{-1}$$

$$= (\tau - e)(\sigma^{-1} - e) + (\tau - e)$$

$$\equiv (\tau - e) \mod I_U I_G.$$

This concludes the proof of Theorem 1.17.

§2. Sylow subgroups

Let G be a finite group of order N. For each prime $p \mid N$ there exists a Sylow subgroup G_p, i.e. a subgroup of order a power of p such that the index $(G : G_p)$ is prime to p. Furthermore, two Sylow subgroups are conjugate.

In particular, if $A \in \mathrm{Mod}(G)$ then $\mathbf{H}^r(G_p, A)$ is well defined, up to a conjugation isomorphism.

By Corollary 1.11 we know that $\mathbf{H}^r(G, A)$ is a torsion group. Therefore

$$\mathbf{H}^r(G, A) = \bigoplus_{p \mid N} \mathbf{H}^r(G, A, p),$$

where $\mathbf{H}^r(G, A, p)$ is the p-primary subgroup of $\mathbf{H}^r(G, A)$, i.e. consists of those elements whose period is a power of p. In particular, if $G = G_p$ is a p-group, then

$$\mathbf{H}^r(G, A) = \mathbf{H}^r(G, A, p).$$

Theorem 2.1. *Let G_p be a p-Sylow subgroup of G. Then for all $r \in \mathbf{Z}$, the restriction*

$$\mathrm{res}^G_{G_p} : \mathbf{H}^r(G, A, p) \to \mathbf{H}^r(G, A)$$

is injective, and the transfer

$$\mathrm{tr}^{G_p}_G : \mathbf{H}^r(G_p, A) \to \mathbf{H}^r(G, A, p)$$

is surjective. We have a direct sum decomposition

$$\mathbf{H}^r(G_p, A) = \mathrm{Im}\ \mathrm{res}^G_{G_p} + \mathrm{Ker}\ \mathrm{tr}^{G_p}_G.$$

Proof. Let $q = (G_p : e)$ be the order of G_p and $m = (G : G_p)$. These integers are relatively prime, and so there exists an integer m' such that $m'm \equiv 1 \mod q$. For all $\alpha \in \mathbf{H}^r(G_p, A)$ we have

$$\alpha = m'm\alpha = m' \cdot \mathrm{tr}^{G_p}_G \mathrm{res}^G_{G_p}(\alpha) = \mathrm{tr}^{G_p}_G m' \cdot \mathrm{res}^G_{G_p}(\alpha),$$

whence the injectivity and surjectivity follow as asserted. For the third, we have for $\beta \in \mathbf{H}^r(G_p, A)$,

$$\beta = \mathrm{res}\ m'\mathrm{tr}(\beta) + (\beta - \mathrm{res}\ m'\mathrm{tr}(\beta)),$$

the restriction and transfer being taken as above. One sees immediately that the first term on the right is the image of the restriction, and the second term is the kernel of the transfer. The sum is direct, because if $\beta = \mathrm{res}(\alpha), \mathrm{tr}(\beta) = 0$, then

$$\mathrm{tr}(\mathrm{res}(\alpha)) = m\alpha = 0,$$

whence $m'm\alpha = \alpha = 0$ and so $\beta = 0$. This concludes the proof.

Corollary 2.2. *Given $r \in \mathbf{Z}$, and $A \in \mathrm{Mod}(G)$, the map*

$$\alpha \mapsto \prod_{p|N} \mathrm{res}^G_{G_p}(\alpha)$$

gives an injective homomorphism

$$\mathbf{H}^r(G, A) \to \prod_{p|N} \mathbf{H}^r(G_p, A).$$

Corollary 2.3. *If $H^r(G_p, A)$ is of finite order for all $p \mid N$, then so is $H^r(G, A)$, and the order of this latter group divides the product of the orders of $H^r(G_p, A)$ for all $p \mid N$.*

Corollary 2.4. *If $H^r(G_p, A) = 0$ for all p, then $H^r(G, A) = 0$.*

§3. Induced representations

Let G be a group and U a subgroup. We are going to define a functor

$$M_G^U : \operatorname{Mod}(U) \longrightarrow \operatorname{Mod}(G).$$

Let $B \in \operatorname{Mod}(U)$. We let $M_G^U(B)$ be the set of mappings from G into B satisfying

$$\sigma f(x) = f(\sigma x) \quad \text{for} \quad \sigma \in U \quad \text{and} \quad x \in G.$$

One can also write

$$M_G^U(B) = (M_G(B))^U.$$

The sum of two mappings is taken as usual summing their values, so $M_G^U(B)$ is an abelian group. We can define an action of G on $M_G^U(B)$ by the formula

$$(\sigma f)(x) = f(x\sigma) \quad \text{for} \quad \sigma, x \in G.$$

We then have transitivity.

Proposition 3.1. *Let $V \subset U$ be subgroups of G. Then the functors*

$$M_G^U \circ M_U^V \quad \text{and} \quad M_G^V$$

are isomorphic in a natural way.

We leave the proof to the reader.

We use the same notation as in §1 with a right coset decomposition $\{c\}$ of U in G, and chosen representatives $\bar{c}$. We continue to use $B \in \operatorname{Mod}(U)$.

Proposition 3.2. *Let G be a group and U a subgroup with right cosets $\{c\}$. Then the map $f \mapsto \operatorname{res} f$, which to an element $f \in M_G^U(B)$ associates its restriction to the coset representatives $\{\bar{c}\}$, is a $\mathbf{Z}$-isomorphism*

$$M_G^U(B) \to M(G/U, B)$$

where $M(G/U, B)$ is the additive group of maps from the coset space G/U into B.

Proof. The formula $f(\sigma\tau) = \sigma f(\tau)$ for $\sigma \in U$ and $\tau \in G$ shows that the values $f(\bar{c})$ of f on coset representatives determine f, so the restriction map above is injective. Furthermore, given a map $f_0 : G/U \to B$, if we define $f_0(\bar{c}) = f_0(c)$, then we may extend f_0 to a map $f : G \to B$ by the same formula, so the proposition is clear.

Proposition 3.3. *Let G be a group and U a subgroup. Then M_G^U is an additive, covariant, exact functor of $\operatorname{Mod}(U)$ to $\operatorname{Mod}(G)$.*

Proof. Let $h : B \to B'$ be a surjective morphism in $\operatorname{Mod}(U)$, and suppose $f' : G/U \to B'$ is a given map. For each value $f'(\bar{c})$ there exists an element $b \in B$ such that $h(b) = f'(\bar{c})$. We may then define a map $f : G/U \to B$ such that $f \circ h = f'$. From this one sees that $M_G^U(B) \to M_G^U(B')$ is surjective. The rest of the proposition is even more routine.

Theorem 3.4. *Let G be a group and U a subgroup. The bifunctors*

$$\operatorname{Hom}_G(A, M_G^U(B)) \quad and \quad \operatorname{Hom}_U(A, B)$$

from $\operatorname{Mod}(G) \times \operatorname{Mod}(U)$ to $\operatorname{Mod}(\mathbf{Z})$ are isomorphic under the following associations. Given $f \in \operatorname{Hom}_G(A, M_G^U(B))$, we let f_1 be the map $A \to B$ such that $f_1(a) = f(a)(e)$. Then f_1 is in $\operatorname{Hom}_U(A, B)$. Conversely, given $h \in \operatorname{Hom}_U(A, B)$ and $a \in A$, let g_a be defined by $g_a(\sigma) = h(\sigma a)$. Then $a \mapsto g_a$ is in $\operatorname{Hom}_G(A, M_G^U(B))$. The maps $f \mapsto f_1, a \mapsto (a \mapsto g_a)$ are inverse to each other.

Proof. Routine verification left to the reader.

The above theorem is fundamental, and is one version of the basic formalism of induced representations. Cf. *Algebra*, Chapter XVIII, §7.

Corollary 3.5. *We have $(M_G^U(B))^G = B^U$.*

Proof. Take $A = \mathbf{Z}$ in the theorem.

Corollary 3.6. *If B is injective in $\mathrm{Mod}(U)$ then $M_G^U(B)$ is injective in $\mathrm{Mod}(G)$.*

Proof. Immediate from the definition of injectivity.

Theorem 3.7. *Let G be a group and U a subgroup. The map*

$$H^r(G, M_G^U(B)) \to H^r(U, B)$$

obtained by composing the restriction res_U^G followed by the U-morphism $g \mapsto g(e)$, is an isomorphism for $r \geq 0$.

Proof. We have two cohomological functors $H_G \circ M_G^U$ and H_U on $\mathrm{Mod}(U)$, because M_G^U is exact. By the two above corollaries, they are both equal to 0 on injective modules, and are isomorphic in dimension 0. By the uniqueness theorem, they are isomorphic in all dimensions. This isomorphism is the one given in the statement of the theorem, because if we denote by $\pi : M_G^U(B) \to B$ the U-morphism such that $\pi g = g(e)$, then

$$H_U(\pi) \circ \mathrm{res}_U^G : H_U \circ M_G^U \to H_U$$

is clearly a δ-morphism which, in dimension 0, induces the prescribed isomorphism $(M_G^U(B))^G$ on B^U. This proves the theorem.

Suppose now that U is of finite index in G. Let $A = M_G^U(B)$. For each coset c, we may define a U-endomorphism $\pi_c : A \to A$ of A into itself by the formula:

$$\pi_c(f)(\sigma) = \begin{cases} 0. & \text{if } \sigma \notin U \\ f(\sigma\bar{c}) & \text{if } \sigma \in U. \end{cases}$$

Indeed, π_c is additive, and if $\tau \in U$, then

$$\tau(\pi_c f)(\sigma) = (\pi_c f)(\tau\sigma) \quad \text{for all} \quad \sigma \in G.$$

Indeed, if $\sigma \notin U$ then both sides are equal to 0; and if $\sigma \in U$, then we use the fact that $f \in M_G^U(B)$ to conclude that they are equal.

Let us denote by A_1 the set of elements $f \in M_G^U(B)$ such that $f(\sigma) = 0$ if $\sigma \notin U$. Then A_1 is a U-module, as one verifies at once.

Theorem 3.8. *Let U be of finite index in G. Then:*

(i) *Let A_1 be the U-submodule of elements $f \in M_G^U(B)$ such that $f(\sigma) = 0$ if $\sigma \notin U$. Then*

$$M_G^U(B) = \bigoplus_c \bar{c}^{-1} A_1,$$

and every such f can be written uniquely in the form

$$f = \sum \bar{c}^{-1}(\pi_c f).$$

(ii) *The map $f \mapsto f(e)$ gives a U-isomorphism $A_1 \xrightarrow{\approx} B$.*

Proof. For the first assertion, let $\sigma = \tau \bar{c}_0$ with $\tau \in U$. Then

$$\sum \bar{c}^{-1}(\pi_c f)(\sigma) = \sum \bar{c}^{-1}(\pi_c f)(\tau \bar{c}_0)$$
$$= \sum (\pi_c f)(\tau \bar{c}_0 \bar{c}^{-1}).$$

If $c \neq c_0$ then the corresponding term is 0. Hence in the above sum, there will be only one term $\neq 0$, with $c = c_0$. In this case, we find the value $f(\tau \bar{c}_0) = f(\sigma)$. This shows that f can be written as asserted, and it is clear that the sum is direct. Finally A_1 is U-isomorphic to B because each $f \mid A_1$ is uniquely determined by its value $f(e)$, taking into account that $\tau f(e) = f(\tau)$ for $\tau \in U$. This same fact shows that we can define $f \mid A_1$ by prescribing $f(e) = b \in B$ and $f(\tau) = \tau b$. This proves the theorem.

We continue to consider the case when U is of finite index in G. Let $A \in \text{Mod}(G)$. We say that A is **semilocal for** U, or **relative to** U, if there exists a U-submodule A_1 of A such that A is equal to the direct sum

$$A = \bigoplus_c \bar{c}^{-1} A_1.$$

We then say that the U-module A_1 is the **local component**. It is clear that A is uniquely determined by its local component, up to an isomorphism. More precisely:

Proposition 3.9. *Let $A_1, A'_1 \in \text{Mod}(U)$.*

(i) *Let $f_1 : A_1 \to A'_1$ be a U-isomorphism, and let A, A' be G-modules, which are semilocal for U, with local components A_1 and A'_1 respectively. Then there exists a unique G-isomorphism $f : A \to A'$ which extends f_1.*

(ii) *Let $A \in \text{Mod}(G)$ and let A_1 be a $\mathbf{Z}$-submodule of A. Suppose that A is direct sum of a finite number of σA_1 (for $\sigma \in G$). Then A is semilocal for the subgroup of elements $\tau \in G$ such that $\tau A_1 = A_1$.*

Proof. Immediate.

Theorem 3.8 and Proposition 3.9 express the fact that to each U-module B there exists a unique G-module semilocal for U, with local component (U, B).

Proposition 3.10. *Let $A \in \text{Mod}(G)$, U a subgroup of finite index in G, and A semilocal for U with local component A_1. Let $\pi_1 : A \to A_1$ be the projection, and π its composition with the inclusion of A_1 in A. Then*

$$1_A = \mathbf{S}_G^U(\pi),$$

in other words, the identity on A is the trace of the projection.

Proof. Every element $a \in A$ can be written uniquely

$$a = \sum_c \bar{c}^{-1} a_c$$

with $a_c \in A_1$. By definition,

$$\mathbf{S}_G^U(\pi)(a) = \sum \bar{c}^{-1} \pi \bar{c} a.$$

The proposition is then clear from the definitions, taking into account the fact that if c, c' are two distinct cosets, then $\pi \bar{c} \bar{c}' a_{c'} = 0$.

If G is finite, one can make the trace more explicit.

Proposition 3.11. *Let G be a group, and U a subgroup of finite index. Let $A \in \mathrm{Mod}(G)$ be semilocal for U with local component A_1. Then an element $a \in A$ is in A^G if and only if*

$$a = \sum_c \bar{c}^{-1} a_1 \quad \text{with some} \quad a_1 \in A_1^U.$$

If G is finite, then $a \in \mathbf{S}_G A$ if and only if $a_1 \in \mathbf{S}_U A_1$ in the above formula. The functors

$$\mathbf{H}_G^0(M_G^U(B)) \qquad \text{and} \qquad \mathbf{H}_U^0(B)$$

(with variable $B \in \mathrm{Mod}(U)$) are isomorphic.

Proof. One verifies at once that for the first assertion, if an element a is expressed as the indicated sum with $a_1 \in A_1$ and $a \in A^G$ then the projection maps A^G into A_1^U. Since we already know that the projection gives an isomorphism between A^G and A_1^U it follows that all elements of A^G are expressed as stated, with $a_1 \in A_1^U$. If G is finite, then for $b \in A$ we have

$$\mathbf{S}_G(b) = \sum_c \bar{c}^{-1} \left(\sum_{\tau \in U} \tau b \right),$$

and the second assertion follows directly from this formula.

For finite groups, M_G^U maps U-regular modules to G-regular modules. This is important because such modules erase cohomology.

Proposition 3.12. *Let G be finite with subgroup U. If A is semilocal for U with local component (U, A_1) and if A_1 is U-regular, then A is G-regular.*

Proof. If one can write $1_{A_1} = \mathbf{S}_U(f)$ with some $\mathbf{Z}$-morphism f, then

$$1_A = \mathbf{S}_G^U(\pi \mathbf{S}_U(f)) = \mathbf{S}_G^U(\mathbf{S}_U(\pi f)) = \mathbf{S}_G(\pi f),$$

which proves that A is G-regular.

From the present view point, we recover a result already found previously.

Corollary 3.13. *Let G be a finite group, with subgroup U and B in $\operatorname{Mod}(U)$. If B is U-regular then $M_G^U(B)$ is G-regular.*

Theorem 3.14. *Let U be a subgroup of finite index in a group G. Suppose $A \in \operatorname{Mod}(G)$ is semilocal for U, with local component A_1. Let $\pi_1 : A \to A_1$ be the projection and $\operatorname{inc} : A_1 \to A$ the inclusion. Then the maps*

$$H_U(\pi_1) \circ \operatorname{res}_U^G \qquad and \qquad \operatorname{tr}_G^U \circ H_U(\operatorname{inc})$$

are inverse isomorphisms

$$H^r(G, A) \xleftrightarrow{\approx} H^r(U, A_1).$$

If G is finite, the same holds for the special functor $\mathbf{H}^r$ instead of H^r.

Proof. The composite

$$A \xrightarrow{\pi_1} A_1 \xrightarrow{\operatorname{inc}} A$$

is a U-morphism of A into itself, which we denoted by π. We know that the identity 1_A is the trace of this morphism. We can then apply Proposition 1.15 to prove the theorem for H. When G is finite, and we deal with the special functor $\mathbf{H}$, we use Corollary 3.13, the uniqueness theorem on cohomological functors vanishing on U-regular modules, the two functors being

$$\mathbf{H}_G \circ M_G^U \qquad and \qquad \mathbf{H}_U.$$

We thus obtain inverse isomorphisms of $\mathbf{H}^r(G, A)$ and $\mathbf{H}^r(U, A_1)$. This concludes the proof.

Remark. Theorem 3.14 is one of the most fundamental of the theory, and is used constantly in algebraic number theory when considering objects associated to a finite Galois extension of a number field.

§4. Double cosets

Let G be a group and U a subgroup of finite index. Let S be an arbitrary subgroup of G. Then there is a disjoint decomposition of G into double cosets

$$G = \bigcup_\gamma U\gamma S = \bigcup_\gamma S\gamma^{-1}U,$$

with $\{\gamma\}$ in some finite subset of G (because U is assumed of finite index), representing the double cosets. For each γ there is a decomposition into simple cosets:

$$S = \bigcup_{\tau_\gamma}(S \cap U[\gamma])\tau_\gamma = \bigcup_{\tau_\gamma} \tau_\gamma^{-1}(S \cap U[\gamma]),$$

where τ_γ ranges over a finite subset of S, depending on γ. Then we claim that the elements $\{\gamma\tau_\gamma\}$ form a family of right coset representatives for U in G, so that

$$G = \bigcup_{\gamma,\tau_\gamma} U\gamma\tau_\gamma = \bigcup_{\gamma,\tau_\gamma} \tau_\gamma^{-1}\gamma_U^{-1}$$

is a decomposition of G into cosets of U. The proof is easy. First, by hypothesis, we have

$$G = \bigcup_{\tau_\gamma}\bigcup_{\gamma} U\gamma(S \cap U[\gamma])\tau_\gamma,$$

and every element of G can be written in the form

$$u_1\gamma\gamma^{-1}u_2\gamma\tau_\gamma = u\gamma\tau_\gamma, \quad \text{with} \quad u_1, u_2 \in U.$$

Second, one sees that the elements $\{\gamma\tau_\gamma\}$ represent different cosets, for if

$$U\gamma\tau_\gamma = U\gamma'\tau_{\gamma'},$$

then $\gamma = \gamma'$ since the γ represent distinct double cosets, whence τ_γ and $\tau\gamma'$ represent the same coset of $\gamma^{-1}U\gamma$, and are consequently equal.

For the rest of this section, we preserve the above notation.

Proposition 4.1. *On the cohomological functor H_U (resp. $\mathbf{H}_U$ if G is finite) on* $\mathrm{Mod}(G)$, *the following morphisms are equal:*

$$\mathrm{res}_S^G \circ \mathrm{tr}_G^U = \sum_\gamma \mathrm{tr}_S^{S\cap U[\gamma]} \circ \mathrm{res}_{S\cap U[\gamma]}^{U[\gamma]} \circ \gamma_*.$$

Proof. As usual, it suffices to verify this formula in dimension 0. Thus let $a \in A^U$. The operation on the left consists in first taking the trace $\mathbf{S}_G^U(a)$, and then applying the restriction which is just the inclusion. The operation on the right consists in taking first $\gamma^{-1}a$, and then applying the restriction which is the inclusion, followed by the trace

$$\mathrm{tr}_S^{S \cap U[\gamma]}(\gamma^{-1}a) = \sum_{\tau_\gamma} \tau_\gamma^{-1}\gamma^{-1}a$$

according to the coset decomposition which has been worked out above. Finally taking the sum over γ, one finds tr_G^U, which proves the proposition.

Corollary 4.2. *If U is normal, then for $A \in \mathrm{Mod}(G)$ and $\alpha \in H^r(U, A)$ we have*

$$\mathrm{res}_U^G \mathrm{tr}_G^U(\alpha) = \mathbf{S}_{G/U}(\alpha),$$

and similarly for **H** *if G is finite.*

Proof. Clear.

Let again $A \in \mathrm{Mod}(G)$ and $\alpha \in H(U, A)$. We say that A is **stable** if for every $\sigma \in G$ we have

$$\mathrm{res}_{U \cap U[\sigma]}^{U[\sigma]} \sigma_*(\alpha) = \mathrm{res}_{U \cap U[\sigma]}^U(\alpha).$$

If U is normal in G, then α is stable if and only if $\sigma_*\alpha = \alpha$ for all $\sigma \in G$.

Proposition 4.3. *Let $\alpha \in H^r(U, A)$ for $A \in \mathrm{Mod}(G)$. If $\alpha = \mathrm{res}_U^G(\beta)$ for some $\beta \in H^r(G, A)$ then α is stable.*

Proof. By Proposition 1.5 we know that $\sigma_*\beta = \beta$. Hence we find

$$\mathrm{res}_{U \cap U[\sigma]}^{U[\sigma]} \circ \sigma_*(\alpha) = \mathrm{res}_{U \cap U[\sigma]}^{U[\sigma]} \circ \sigma_* \circ \mathrm{res}_U^G(\beta)$$

$$= \mathrm{res}_{U \cap U[\sigma]}^{U[\sigma]} \mathrm{res}_{U[\sigma]}^G \sigma_*(\beta)$$

$$= \mathrm{res}_{U \cap U[\sigma]}^G(\beta).$$

If we unwind this formula via the intermediate subgroup U instead of $U[\sigma]$, we find what we want to prove the proposition.

Proposition 4.4. *If $\alpha \in H^r(U, A)$ is stable, then*

$$\operatorname{res}_U^G \circ \operatorname{tr}_G^U(\alpha) = (G : U)\alpha.$$

Proof. We apply the general formula to the case when $U = S$. The verification is immediate.

Suppose finally that A is semilocal for U, with local component A_1, projection $\pi_1 : A \to A_1$ as before. Then elements of S possibly do not permute the submodules A_c transitively. However, we have:

Proposition 4.5. *Let $\sigma \in G$. Then $\sigma \in S\gamma^{-1}U$ if and only if σA_1 is in the same orbit of S as $\gamma^{-1}A_1$. For each γ, the sum*

$$\sum_{\tau_\gamma} \tau_\gamma^{-1} \gamma^{-1} A_1$$

is an S-module, semilocal for $S \cap U[\gamma]$, with local component $\tau_\gamma^{-1} \gamma^{-1} A_1$.

Proof. The assertion is immediate from the decomposition of G into cosets $\tau_\gamma^{-1} \gamma^{-1} U$.

If S and U are both of finite index in G, then the reader will observe a symmetry in the above formulas, in particular in the double coset decomposition. In particular, we can rewrite the formula of Proposition 4.3 in the form

$$(*) \qquad \operatorname{res}_U^G \circ \operatorname{tr}_G^S = \sum_\gamma \gamma_* \circ \operatorname{tr}_{U[\gamma]}^{U[\gamma] \cap S} \circ \operatorname{res}_{U[\gamma] \cap S}^S.$$

We just take into account the commutativity of γ_* and the other maps, replacing U by S, S by U and γ by γ^{-1}.

CHAPTER III
Cohomological Triviality

In this chapter we consider only finite groups, and the special functor $\mathbf{H}_G$, such that $\mathbf{H}_G(A) = A^G/\mathbf{S}_G A$. The main result is Theorem 1.7.

§1. The twins theorem

We begin by auxiliary results. We let $\mathbf{F}_p = \mathbf{Z}/p\mathbf{Z}$ for a prime p. We always assume G has trivial action on $\mathbf{F}_p$.

Proposition 1.1. *Let G be a p-group, and $A \in \mathrm{Mod}(G)$ finite of order equal to a p-power. Then $A^G = 0$ implies $A = 0$.*

Proof. We express A as a disjoint union of orbits of G. For each $x \in A$ we let G_x be the isotropy group, i.e. the subgroup of elements $\sigma \in G$ such that $\sigma x = x$. Then the number of elements in the orbit Gx is the index $(G : G_x)$. Since G leaves 0 fixed, and

$$(A : 0) = \sum m_i p^i$$

where m_i is the number of orbits having p^i elements, it follows that either $A = 0$ or there is an element $x \neq 0$ whose orbit also has only one element, i.e. x is fixed by G, as was to be shown.

Corollary 1.2. *Let G be a p-group. If A is a simple p-torsion G-module then $A \approx \mathbf{F}_p$.*

Proof. Immediate.

Corollary 1.3. *The radical of $\mathbf{F}_p[G]$ is equal to the ideal I_p generated by all elements $(\sigma - e)$ over $\mathbf{F}_p$.*

Proof. A simple G-module over $\mathbf{F}_p[G]$ is finite, of order a power of p, so isomorphic to $\mathbf{F}_p$, annihilated by I_p which is therefore contained in the radical. The reverse inclusion is immediate since $\mathbf{F}_p[G]/I_p \approx \mathbf{F}_p$.

Proposition 1.4. *Let G be a p-group and A an $\mathbf{F}_p[G]$-module. The following conditions are equivalent.*

1. $\mathbf{H}^i(G, A) = 0$ for some i with $-\infty < i < \infty$.

2. A is G-regular.

3. A is G-free.

Proof. Since $\mathbf{F}_p$ is a field every $\mathbf{F}_p$-module is free in $\mathrm{Mod}(\mathbf{F}_p)$. If A is G-free then A is obviously G-regular. Conversely, if A is G-regular, with local component A_1 for the unit element of G, then A_1 is a direct sum of $\mathbf{F}_p$ a certain number of times, and the G-orbit of a factor $\mathbf{F}_p$ is G-isomorphic to $\mathbf{F}_p[G]$, so A itself is a G-direct sum of such G-modules isomorphic to $\mathbf{F}_p[G]$, thus proving the equivalence of the last two conditions.

It suffices now to prove the equivalence between the first and third conditions. Abbreviate $I = I_p$ for the augmentation ideal of $\mathbf{F}_p[G]$. Then A/IA is a vector space over $\mathbf{F}_p$. Let $\{a_j\}$ be representatives in A for a basis over $\mathbf{F}_p$, so that A is generated over $\mathbf{F}_p[G]$ by these elements a_j and IA. Let E be the $\mathbf{F}_p[G]$-free module with free generators $\bar{a}_j$. There is a G-morphism $E \to A$ such that $\bar{a}_j \mapsto a_j$. Let B be the image of E in A, so that $A = B + IA$. Since I is nilpotent, $I^n = 0$ for some positive integer n, and we find

$$A = B + IA + B + IB + I^2 A = \cdots = B + IB + \cdots + I^n A = B$$

by iteration. Hence the map $E \to A$ is surjective. Let A' be its kernel. We have

$$E/IE = A/IA$$

and so $A' \subset IE$. Hence $\mathbf{S}_G A' = 0$. Since the following sequence is exact,

$$0 \to A' \to E \to A \to 0,$$

and E is G-regular, one finds $\mathbf{H}^r(G, A) = \mathbf{H}^{r+1}(G, A')$ for all $r \in \mathbf{Z}$.

Suppose that the index i in the hypothesis is -2. Then $\mathbf{H}^{-1}(G, A') = 0$, so $A'_{\mathbf{S}_G} = IA'$. Since $A' = A'_{\mathbf{S}_G}$, we find $A' = IA'$ and hence $A' = 0$ so $E \approx A$.

On the other hand, if $i \neq -2$, we know by dimension shifting that there exists a G-module C such that $\mathbf{H}^r(U, C) \approx \mathbf{H}^{r+1}(U, A)$ for some integer d, all $r \in \mathbf{Z}$ and all subgroups U of G, and also $\mathbf{H}^{-2}(G, C) = 0$. Hence C is $\mathbf{F}_p[G]$-free, so cohomologically trivial, and therefore similarly for A, so in particular $\mathbf{H}^{-2}(G, A) = 0$. This proves the proposition.

Proposition 1.5. *Let G be a p-group, $A \in \mathrm{Mod}(G)$. Suppose there exists an integer i such that $\mathbf{H}^i(G, A) = \mathbf{H}^{i+1}(G, A) = 0$. Suppose in addition that A is $\mathbf{Z}$-free. Then A is G-regular.*

Proof. We have an exact sequence

$$0 \to A \xrightarrow{p} A \to A/pA \to 0$$

and hence

$$0 = \mathbf{H}^i(A) \to \mathbf{H}^i(A/pA) \to \mathbf{H}^{i+1}(A) = 0,$$

the functor $\mathbf{H}$ being $\mathbf{H}_G$. Therefore $\mathbf{H}^i(A/pA) = 0$. Since A/pA is an $\mathbf{F}_p[G]$-module, we conclude from Proposition 1.3 that A/pA is $\mathbf{F}_p[G]$-free, and therefore G-regular.

Since we supposed A is $\mathbf{Z}$-free, we see immediately that the sequence

$$0 \to \mathrm{Hom}_A(A, A) \xrightarrow{p} \mathrm{Hom}_A(A, A) \to \mathrm{Hom}_{\mathbf{Z}}(A, A/pA) \to 0$$

is exact. But $\mathrm{Hom}_{\mathbf{Z}}(A, A/pA)$ is G-regular, so

$$p = p_* : \mathbf{H}^i(\mathrm{Hom}_{\mathbf{Z}}(A, A)) \to \mathbf{H}^i(\mathrm{Hom}_{\mathbf{Z}}(A, A))$$

is an automorphism. Hence so is its iteration, and hence so is multiplication by the order $(G : e)$ since G is assumed to be a p-group. But $(G : e)_* = 0$, whence all the cohomology groups $\mathbf{H}^r(\mathrm{Hom}_{\mathbf{Z}}(A, A)) = 0$ for $r \in \mathbf{Z}$. In particular, $\mathbf{H}^0(\mathrm{Hom}_A(A, A)) = 0$. From the definitions, we conclude that the identity 1_A is a trace, and so A is G-regular, thus proving Proposition 1.5.

Corollary 1.6. *Hypotheses being as in the proposition, then A is projective in* $\mathrm{Mod}(G)$.

Proof. Immediate consequence of Chapter I, Proposition 2.13.

Let G be a finite group and $A \in \mathrm{Mod}(G)$. We define A to be **cohomologically trivial** if $\mathbf{H}^r(U, A) = 0$ for all subgroups U of G and all $r \in \mathbf{Z}$.

Theorem 1.7. Twin theorem. *Let G be a finite group and $A \in \mathrm{Mod}(G)$. Then A is cohomologically trivial if and only if for each $p \mid (G : e)$ there exists an integer i_p such that*

$$\mathbf{H}^{i_p}(G_p, A) = \mathbf{H}^{i_p+1}(G_p, A) = 0$$

for a p-Sylow subgroup G_p of G.

Proof. Let $E \in \mathbf{Z}[G]$ be G-free such that

$$0 \to A' \to E \to A \to 0$$

is exact. Then

$$\mathbf{H}^{i_p+1}(G_p, A) = \mathbf{H}^{i_p+2}(G_p, A') = 0.$$

Since A' is $\mathbf{Z}$-free, it is also G_p-regular by Proposition 1.5. Hence for all subgroups G'_p of G_p we have $\mathbf{H}^r(G'_p, A) = 0$ for all $r \in \mathbf{Z}$. Since there is an injection

$$0 \to \mathbf{H}^r(G', A) \to \prod_p \mathbf{H}^r(G'_p, A)$$

for all subgroups G' of G by Chapter II, Corollary 2.2. It follows that A is cohomologically trivial. The converse is obvious.

Corollary 1.8. *Let G be finite and $A \in \mathrm{Mod}(G)$. The following conditions are equivalent:*

1. *A is cohomologically trivial.*

2. *The projective dimension of A is $\leqq 2$.*

3. *The projective dimension of A is finite.*

Proof. Recall from *Algebra* that A has projective dimension $\leqq s < \infty$ if one can find an exact sequence

$$0 \to P_1 \to P_2 \to \cdots \to P_s \to A \to 0$$

with projectives P_j. One can complete this sequence by introducing the kernels and cokernels as shown, the arches being exact.

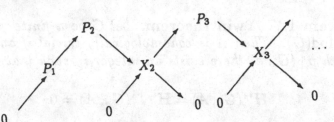

Therefore

$$\mathbf{H}^r(G', A) = \mathbf{H}^{r+1}(G', X_{s-1}) = \cdots = \mathbf{H}^{r+s-1}(G', P_1) = 0.$$

It is clear that a G-module of finite projective dimension is cohomologically trivial.

Conversely, let us write an exact sequence

$$0 \to A' \to P \to A \to 0$$

where P is $\mathbf{Z}[G]$-free. Then A' is $\mathbf{Z}$-free, and by Proposition 1.5 it is also G_p-regular for all p. We now need a lemma.

Lemma 1.9. *Suppose $M \in \mathrm{Mod}(G)$ is $\mathbf{Z}$-free and G_p-regular for all primes p. Then M is G-regular, and so projective in $\mathrm{Mod}(G)$.*

Proof. We view $\mathbf{H}^0(G, \mathrm{Hom}_{\mathbf{Z}}(M, M))$ as being injected in the product

$$\prod_p \mathbf{H}^0(G_p, \mathrm{Hom}_{\mathbf{Z}}(M, M)),$$

and we apply the definition. We conclude that M is G-regular. Since M is $\mathbf{Z}$-free it is $\mathbf{Z}[G]$-projective by Corollary 1.6.

We apply the lemma to $M = A'$ to conclude that the projective dimension of A is $\leqq 2$. This proves Corollary 1.8.

Corollary 1.10. *Let $A \in \text{Mod}(G)$ be cohomologically trivial, and let $M \in \text{Mod}(G)$ be without torsion. Then $A \otimes M$ is cohomologically trivial.*

Proof. There is an exact sequence

$$0 \to P_1 \to P_2 \to A \to 0$$

with P_1, P_2 projective in $\text{Mod}(G)$. Since M has no torsion, the sequence

$$0 \to P_1 \otimes M \to P_2 \otimes M \to A \otimes M \to 0$$

is exact. But P_1, P_2 are G-regular (direct summands in free modules, so regular), whence $P_i \otimes M$ is cohomologically trivial for $i = 1, 2$, whence $A \otimes M$ is cohomologically trivial.

More generally, we have one more result, which we won't use in the sequel.

Suppose A is cohomologically trivial, and that we have an exact sequence

$$0 \to P_1 \to P_2 \to A \to 0$$

with projectives P_1, P_2. Then we have an exact sequence

$$\to \text{Tor}^1(P_2, M) \to \text{Tor}^1(A, M) \to P_1 \otimes M \to P_2 \otimes M \to A \otimes M \to 0$$

for an arbitrary $M \in \text{Mod}(G)$. Furthermore $\text{Tor}^1(P_2, M) = 0$ since P_2 has no torsion (because $\mathbf{Z}[G]$ is $\mathbf{Z}$-free). By dimension shifting, and similar reasoning for Hom, we find:

Theorem 1.11. *Let G be a finite group, $A, B \in \text{Mod}(G)$, and suppose A or B is cohomologically trivial. Then for all $r \in \mathbf{Z}$ and all subgroups G' of G, we have*

$$\mathbf{H}^r(G', A \otimes B) \approx \mathbf{H}^{r+2}(G', \text{Tor}_1^{\mathbf{Z}}(A, B))$$

$$\mathbf{H}^r(G', \text{Hom}(A, B)) \approx \mathbf{H}^{r-2}(G', \text{Ext}_{\mathbf{Z}}^1(A, B)).$$

Corollary 1.12. *Let G, A, B be as in the theorem. Then $A \otimes B$ is cohomologically trivial if and only if $\text{Tor}_1^{\mathbf{Z}}(A, B)$ is cohomologically trivial; and $\text{Hom}(A, B)$ is cohomologically trivial if and only if $\text{Ext}_{\mathbf{Z}}^1(A, B)$ is cohomologically trivial.*

Corollary 1.13. *Let G, A, B be as in the theorem. Then $A \otimes B$ is cohomologically trivial if A or B is without p-torsion for each prime p dividing $(G : e)$.*

§2. The triplets theorem

Let $f : A \to B$ be a morphism in $\mathrm{Mod}(G)$. Let U be a subgroup of G, and let

$$f_r : \mathbf{H}^r(U, A) \to \mathbf{H}^r(U, B)$$

be the homomorphisms induced on cohomology. Actually we should write $f_{r,U}$ but we omit the index U for simplicity. We say that f is a **cohomology isomorphism** if f_r is an isomorphism for all r and all subgroups U. We say that A and B are **cohomologically equivalent** if there exists a cohomology isomorphism f as above.

Theorem 2.1. *Let $f : A \to B$ be a morphism in $\mathrm{Mod}(G)$, and suppose there exists some $i \in \mathbf{Z}$ such that f_{i-1} is surjective, f_i is an isomorphism, and f_{i+1} is injective, for all subgroups U of G. Then f is a cohomology isomorphism.*

Proof. Suppose first that f is injective. We shall reduce the general case to this special case later. We therefore have an exact sequence

$$0 \to A \xrightarrow{f} B \xrightarrow{g} C \to 0,$$

with $C = B/fA$, and the corresponding cohomology sequence

$$\longrightarrow \mathbf{H}^{i-1}(U,A) \xrightarrow{f_{i-1}} \mathbf{H}^{i-1}(U,B) \xrightarrow{g_{i-1}} \mathbf{H}^{i-1}(U,C) \longrightarrow$$

$$\xrightarrow{\delta_i} \mathbf{H}^i(U,A) \xrightarrow{f_i} \mathbf{H}^i(U,B) \xrightarrow{g_i} \mathbf{H}^i(U,C) \longrightarrow$$

$$\xrightarrow{\delta_{i+1}} \mathbf{H}^{i+1}(U,A) \xrightarrow{f_{i+1}} \mathbf{H}^i(U,B) \longrightarrow$$

We shall see that $\mathbf{H}^{i-1}(U, C) = \mathbf{H}^i(U, C) = 0$. As to $\mathbf{H}^{i-1}(U, C) = 0$, it comes from the fact that f_{i-1} surjective implies $g_{i-1} = 0$, and f_i being an isomorphism implies $\delta_{i-1} = 0$. As to $\mathbf{H}^i(U, C) = 0$, it comes from the fact that f_i surjective implies $g_i = 0$, and f_{i+1} injective implies $\delta_{i+1} = 0$. By the twin theorem, we conclude that $\mathbf{H}^r(U, C) = 0$ for all $r \in \mathbf{Z}$, whence f_r is an isomorphism for all r.

We now reduce the theorem to the preceding case, by the method of the mapping cylinder. Let us put $\mathbf{M}_G(A) = \bar{A}$ and $\varepsilon_A = \varepsilon$. We

have an injection
$$\varepsilon : A \to \bar{A}.$$

We map A into the direct sum $B \oplus \bar{A}$ by
$$\bar{f} : A \to B \oplus \bar{A} \quad \text{such that} \quad \bar{f}(a) = f(a) + \varepsilon(a).$$

One sees at once that $\bar{f}$ is a morphism in $\mathrm{Mod}(G)$. We have an exact sequence

$$0 \to A \xrightarrow{\bar{f}} B \oplus \bar{A} \xrightarrow{h} C \to 0$$

where C is the cokernel of $\bar{f}$.

We also have the projection morphism

$$p : B \oplus \bar{A} \to B \quad \text{defined by} \quad p(b + \bar{a}) = b.$$

Its kernel is $\bar{A}$, and we have $f = p\bar{f}$, whence the commutative diagram:

$$
\begin{array}{ccccccccc}
& & & & 0 & & & & \\
& & & & \downarrow & & & & \\
0 & \longrightarrow & A & \xrightarrow{\bar{f}} & B \oplus \bar{A} & \longrightarrow & C & \longrightarrow & 0 \\
& & & \searrow & \downarrow & & & & \\
& & & & B & & & &
\end{array}
$$

We then obtain the diagram

$$\mathbf{H}^i(\bar{A}) = 0$$
$$\downarrow$$
$$\mathbf{H}^{i-1}(C) \longrightarrow \mathbf{H}^i(A) \xrightarrow{f_i} \mathbf{H}^i(B \oplus \bar{A}) \xrightarrow{h_i} \mathbf{H}^i(C) \longrightarrow \mathbf{H}^{i+1}(A)$$

$$f_i \qquad\qquad \downarrow p_i \qquad\qquad g_i$$

$$\mathbf{H}^i(B)$$

$$\downarrow$$

$$\mathbf{H}^{i+1}(\bar{A}) = 0$$

the cohomology groups $\mathbf{H}$ being $\mathbf{H}_U$ for any subgroup U of G. The triangles are commutative.

The extreme vertical maps in the middle are 0 because $\bar{A} = \mathbf{M}_G(A)$, and consequently p_i is an isomorphism, which has an inverse p_i^{-1}. We have put

$$g_i = h_i p_i^{-1}$$

From the formula $f = p\bar{f}$ we obtain $f_i = p_i \bar{f}_i$. We can therefore replace $\mathbf{H}^i(B \oplus \bar{A})$ by $\mathbf{H}^i(B)$ in the horizontal sequence, and we obtain an exact sequence which is the same as the one obtained in the first part of the proof. Thus the theorem is reduced to this first part, thus concluding the proof.

§3. The splitting module and Tate's theorem

The second cohomology group in many cases, especially class field theory, plays a particularly important role. We shall describe here a method to kill a cocycle in dimension 2.

Let G have order n and let $\alpha \in \mathbf{H}^2(G, A)$. Recall the exact sequence

$$0 \to I_G \to \mathbf{Z}[G] \to \mathbf{Z} \to 0,$$

which is $\mathbf{Z}$-split, and induces an isomorphism

$$\delta : H^r(G, \mathbf{Z}) \to H^{r+1}(G, I_G) \quad \text{for all} \quad r.$$

Theorem 3.1. Let $A \in \mathrm{Mod}(G)$ and $\alpha \in H^2(G, A)$. There exists $A' \in \mathrm{Mod}(G)$ and an exact sequence

$$0 \to A \xrightarrow{u} A' \to I_G \to 0,$$

splitting over $\mathbf{Z}$, such that $\alpha = \delta\delta\zeta$, where ζ is the generator of $\mathbf{H}^0(G, \mathbf{Z})$ corresponding to the class of 1 in $H^0(G, \mathbf{Z}) = \mathbf{Z}/n\mathbf{Z}$, and

$$u_*\alpha = 0,$$

in other words, α splits in A'.

Proof. We define A' to be the direct sum of A and a free abelian group on elements $x_\sigma (\sigma \in G, \sigma \neq e)$. We define an action of G on

A' by means of a cocycle $\{a_{\sigma,\tau}\}$ representing α. We put $x_e = a_{e,e}$ for convenience, and let

$$\sigma x_\tau = x_{\sigma\tau} - x_\sigma + a_{\sigma,\tau}.$$

One verifies by brute force that this definition is consistent by using the coboundary relation satisfied by the cocycle $\{a_{\sigma,\tau}\}$, namely

$$\lambda a_{\sigma,\tau} - a_{\lambda\sigma,\tau} + a_{\lambda,\sigma\tau} - a_{\sigma,\tau} = 0.$$

One sees trivially that α splits in A'. Indeed, $(a_{\sigma,\tau})$ is the coboundary of the cochain (x_σ).

We define a morphism $v : A' \to I_G$ by letting

$$v(a) = 0 \text{ for } a \in A \text{ and } v(x_\sigma) = \sigma - e \text{ for } \sigma \neq e.$$

Here we identify A as a direct summand of A'. The map v is a G-morphism in light of the definition of the action of G on A'. It is obviously surjective, and the kernel of v is equal to A.

There remains to verify that $\alpha = \delta\delta\zeta$. The coboundary $\delta\zeta$ is represented by the 1-cocycle $b_\sigma = \sigma - e$ in I_G, representing an element β of $\mathbf{H}^1(G, I_G)$. We find $\delta\beta$ by selecting a cochain of G in A', for instance (x_σ), such that $v(x_\sigma) = b_\sigma$. The coboundary of (x_σ) represents $\delta\beta$, and one then sees that this gives α, thus proving the theorem.

We define an element $A \in \mathrm{Mod}(G)$ to be a **class module** if for every subgroup U of G, we have $\mathbf{H}^1(U, A) = 0$, and if $\mathbf{H}^2(G, A)$ is cyclic of order $(G : e)$, generated by an element α such that $\mathrm{res}_U^G(\alpha)$ generates $\mathbf{H}^2(U, A)$ and is of order $(U : e)$. An element α as in this definition will be called **fundamental**. The terminology comes from class field theory, where one meets such modules. See also Chapter IX.

Theorem 3.2. *Let $G, A, \alpha, u : A \to A'$ be as in Theorem 3.1. Then A is a class module and α is fundamental if and only if A' is cohomologically trivial.*

Proof. Suppose A is a class module and α fundamental. We have an exact sequence for all subgroups U of G:

$$0 \to \mathbf{H}^1(A) \to \mathbf{H}^1(A') \to \mathbf{H}^1(I) \to \mathbf{H}^2(A) \to \mathbf{H}^2(A') \to 0.$$

The 0 furthest to the right is due to the fact that

$$\mathbf{H}^2(I_G) = \mathbf{H}^1(\mathbf{Z}) = 0.$$

Since

$$\alpha = \delta\beta \text{ and } \beta = \delta\zeta,$$

and $\mathbf{H}^1(U, I)$ is cyclic of order $(U : e)$ generated by β, it follows that

$$\mathbf{H}^1(I) \to \mathbf{H}^2(A)$$

is an isomorphism. We conclude that $\mathbf{H}^1(U, A') = 0$ for all U.

Since we also have the exact sequence

$$\mathbf{H}^2(A) \to \mathbf{H}^2(A') \to \mathbf{H}^2(I) = 0,$$

and α splits in A', we conclude that $\mathbf{H}^2(U, A') = 0$ for all U. Hence A' is cohomologically trivial by the twin theorem.

Conversely, suppose A' cohomologically trivial. Then we have isomorphisms

$$\mathbf{H}^1(I) \xrightarrow{\delta} \mathbf{H}^2(A) \quad \text{and} \quad \mathbf{H}^0(\mathbf{Z}) \xrightarrow{\delta} \mathbf{H}^1(I),$$

for all subgroups U of G. This shows that $\mathbf{H}^2(U, A)$ is indeed cyclic of order $(U : e)$, generated by $\delta\delta\zeta$. This concludes the proof.

CHAPTER IV
Cup Products

§1. Erasability and uniqueness

To treat cup products, we have to start with the general notion of multilinear categories, due to Cartier.

Let $\mathfrak{A}$ be an abelian category. A structure of **multilinear category** on $\mathfrak{A}$ consists in being given, for each $(n+1)$-tuple $A_1, \ldots, A_n, B$ of objects in $\mathfrak{A}$, an abelian group $L(A_1, \ldots, A_n, B)$ satisfying the following conditions.

MUL 1. For $n = 1, L(A, B) = \operatorname{Hom}(A, B)$.

MUL 2. Let

$$f_1 : \quad A_{11} \times \cdots \times A_{ln_1} \quad \to \quad B_1$$
$$\cdots$$
$$f_r : \quad A_{r1} \times \cdots \times A_{rn_r} \quad \to \quad B_r$$
$$g : \qquad B_1 \times \cdots \times B_r \qquad \to \quad C$$

be multilinear. Then we may compose $g(f_1, \ldots, f_r)$ in $L(A_{11}, \ldots, A_{rn_r}, C)$, and this composition is multilinear in $g, f_1, \ldots, f_r$.

MUL 3. With the same notation, we have $g(\operatorname{id}, \ldots, \operatorname{id}) = g$.

MUL 4. The composition is associative, in the sense that (with obvious notation)

$$g(f_1(h\ldots),f_2(h\ldots),\ldots,f_r(h\ldots)) = g(f_1,\ldots,f_r)(h\ldots).$$

As usual the reader may think in terms of ordinary multilinear maps on abelian groups. These define a multilinear category, from which others can be defined by placing suitable conditions.

Example. Let G be a group. Then $\mathrm{Mod}(G)$ is a multilinear category if we define $L(A_1,\ldots,A_n,B)$ to consist of those $\mathbf{Z}$-multilinear maps θ satisfying

$$\theta(\sigma a_1,\ldots,\sigma a_n) = \sigma\theta(a_1,\ldots,a_n)$$

for all $\sigma \in G$ and $a_i \in A_i$.

We can extend in the obvious way the notion of functor to multilinear categories. Explicitly, a functor $F : \mathfrak{A} \to \mathfrak{B}$ of such a category into another is given by a map $T : f \mapsto T(f) = f_*$, which to each multilinear map f in $\mathfrak{A}$ associates a multilinear map in B, satisfying the following condition. Let

$$f_1 : A_1 \quad \times \cdots \times \quad A_{N_1} \longrightarrow B_1$$
$$\cdots\cdots$$
$$f_p : A_{n_{p-1}} \times \cdots \times \quad A_{n_p} \longrightarrow B_p$$
$$g : B_1 \quad \times \cdots \times \quad B_p \longrightarrow C$$

be multilinear in $\mathfrak{A}$. Then we can compose $g(f_1,\ldots,f_p)$ and $Tg(Tf_1,\ldots,Tf_p)$. The condition is that

$$T(g(f_1,\ldots,f_p)) = Tg(Tf_1,\ldots,Tf_p) \quad \text{and} \quad T(\mathrm{id}) = \mathrm{id}.$$

We could also define the notion of tensor product in a multilinear abelian category. It is a bifunctor, bilinear, on $\mathfrak{A} \times \mathfrak{A}$, satisfying the universal mapping property just as for the ordinary tensor product. In the applications, it will be made explicit in each case how such a tensor product arises from the usual one.

Furthermore, in the specific cases of multilinear abelian categories to be considered, the category will be closed under taking tensor products, i.e. if A, B are objects of $\mathfrak{A}$, then the linear factorization of a multilinear map is also in $\mathfrak{A}$.

Let now $E_1 = (E_1^{p_1}), \ldots, E_n = (E_n^{p_n})$ and $H = (H^r)$ be δ-functors on the abelian category $\mathfrak{A}$, which we suppose multilinear, and the functors have values in a multilinear abelian category $\mathfrak{B}$. We assume that for each value of $p_1, \ldots, p_n$ taken on by $E_1, \ldots, E_n$ respectively, the sum $p_1 + \cdots + p_n$ is among the values taken on by r. By a **cup product**, or **cupping**, of $E_1, \ldots, E_n$ into H we mean that to each multilinear map $\theta \in L(A_1, \ldots, A_n; B)$ we have associated a multilinear map

$$\theta_{(p)} = \theta_* = \theta_{p_1, \ldots, p_n} : E_1^{p_1}(A_1) \times \cdots \times E_n^{p_n}(A_n) \to H^p(B)$$

where $p = p_1 + \cdots + p_n$, satisfying the following conditions.

Cup 1. The association $\theta \mapsto \theta_{(p)}$ is a functor from the multilinear category $\mathfrak{A}$ into $\mathfrak{B}$ for each $(p) = (p_1, \ldots, p_n)$.

Cup 2. Given exact sequences in $\mathfrak{A}$,

$$0 \to A_i' \to A_i \to A_i'' \to 0$$

$$0 \to B' \to B \to B'' \to 0$$

and multilinear maps f', f', f'' in $\mathfrak{A}$ making the following diagram commutative:

$$A_1 \times \cdots \times A_i' \times \cdots \times A_n \to A_1 \times \cdots \times A_i \times \cdots \times A_n \to A_1 \times \cdots \times A_i'' \times \cdots \times A_n$$
$$\downarrow f' \qquad\qquad\qquad \downarrow f \qquad\qquad\qquad \downarrow f''$$
$$B' \qquad \to \qquad B \qquad \to \qquad B''$$

then the diagram

$$
\begin{array}{ccc}
E_1^{p_1}(A_1) \times \cdots \times E_i^{p_i}(A_i'') \times \cdots \times E_n^{p_n}(A_n) & \xrightarrow{f_*''} & H^p(B'') \\
\text{id} \downarrow \qquad\qquad \delta \downarrow \qquad\qquad \text{id} \downarrow & & \downarrow \delta \\
E_1^{p_1}(A_1) \times \cdots \times E_i^{p_i+1}(A_i') \times \cdots \times E_n^{p_n}(A_n) & \xrightarrow{f_*'} & H^{p+1}(B')
\end{array}
$$

has character $(-1)^{p_1 + \cdots + p_{i-1}}$, which means

$$f_*'(\text{id}, \ldots, \delta, \ldots, \text{id}) = (-1)^{p_1 + \cdots + p_{i-1}} \delta \circ f_*''.$$

The accumulation of indices is inevitable if one wants to take all possibilities into account. In practice, we mostly have to deal with the following cases.

First, let H be a cohomological functor. For each $n \geq 1$ suppose given a cupping

$$H \times \cdots \times H \to H$$

(the product on the left occurring n times) such that for $n = 1$ the cupping is the identity. Then we say that H is a **cohomological cup functor**.

Next, suppose we have only two factors, i.e. a cupping

$$E \times F \to H$$

from two δ-functors into another. Most of the time, instead of indexing the induced maps by their degrees, we simply index them by a $*$.

If $\mathfrak{A}$ is closed under the tensor product, then by **Cup 1** a cohomological cup functor is uniquely determined by its values on the canonical bilinear maps

$$A \times B \to A \otimes B.$$

Indeed, if $f : A \times B \to C$ is bilinear, one can factorize f through $A \otimes B$,

$$A \times B \xrightarrow{\theta} A \otimes B \xrightarrow{\varphi} C$$

where θ is bilinear and φ is a morphism in $\mathfrak{A}$. Thus certain theorems will be reduced to the study of the cupping on tensor products.

Let $E = (E^p)$ and $F = (F^q)$ be δ-functors with a cupping into the δ-functor $H = (H^r)$. Given a bilinear map

$$A \times B \to C$$

and two exact sequences

$$0 \longrightarrow A' \longrightarrow A \longrightarrow A'' \longrightarrow 0$$

$$0 \longrightarrow C' \longrightarrow C \longrightarrow C'' \longrightarrow 0$$

such that the bilinear map $A \times B \to C$ induces bilinear maps

$$A' \times B \to C' \quad \text{and} \quad A'' \times B \to C'',$$

then we obtain a commutative diagram

$$
\begin{array}{ccc}
E^p(A'') \ \times \ \ F^q(B) & \longrightarrow & H^{p+q}(C'') \\
\downarrow \delta & \downarrow & \downarrow \delta \\
E^{p+1}(A') \ \times \ \ F^q(B) & \longrightarrow & H^{p+q+1}(C')
\end{array}
$$

and therefore we get the formula

$$\delta(\alpha''\beta) = (\delta\alpha'')\beta \quad \text{with} \quad \alpha'' \in E^p(A'') \quad \text{and} \quad \beta \in F^q(B).$$

Proposition 1.1. *Let H be a cohomological cup functor on a multilinear category $\mathfrak{A}$. Then the product*

$$(\alpha, \beta) \mapsto (-1)^{pq}\beta\alpha \qquad for \qquad \alpha \in H^p(A), \beta \in H^q(B)$$

also defines a cupping $H \times H \to H$, making H into another cohomological cup functor, equal to the first one in dimension 0.

Proof. Clear.

Remark. Since we shall prove a uniqueness theorem below, the preceding proposition will show that we have

$$\alpha\beta = (-1)^{pq}\beta\alpha.$$

Now we come to the question of uniqueness for cup products. It will be applied to the uniqueness of a cupping on a cohomological functor H given in one dimension. More precisely, let H be a cohomological cup functor on a multilinear category $\mathfrak{A}$. When we speak of the cup functor in dimension 0, we mean the cupping

$$H^0 \times H^0 \times \cdots \times H^0 \to H^0,$$

which to each multilinear map $\theta : A_1 \times \cdots \times A_n \to B$ associates the multilinear map

$$\theta_0 : H^0(A_1) \times \cdots \times H^0(A_n) \to H^0(B).$$

We note that to prove a uniqueness theorem, we only need to deal with two factors (i.e. bilinear morphisms), because a cupping of several functors can be expressed in terms of cuppings of two functors, by associativity.

Theorem 1.2. *Let $\mathfrak{A}$ be an abelian multilinear category. Let $E = (E^0, E^1)$ and $H = (H^0, H^1)$ be two exact δ-functors, and let F be a functor of $\mathfrak{A}$ into a multilinear category $\mathfrak{B}$. Suppose E^1 is erasable by (M, ε), that $\mathfrak{A}, \mathfrak{B}$ are closed under tensor products, and that for all $A, B \in \mathfrak{A}$ the sequence*

$$0 \to A \otimes B \xrightarrow{\varepsilon_A \otimes 1} M_A \otimes B \to X_A \otimes B \to 0$$

is exact. Suppose given a cupping $E \times F \to H$. Then we have a commutative diagram

$$
\begin{array}{ccccc}
E^0(X_A) & \times & F(B) & \to & H^0(A \otimes B) \\
\downarrow{\scriptstyle \delta} & & \downarrow{\scriptstyle \mathrm{id}} & & \downarrow{\scriptstyle \delta} \\
E^1(A) & \times & F(B) & \to & H^1(A \otimes B)
\end{array}
$$

and the coboundary on the left is surjective.

Proof. Clear.

Corollary 1.3. *Hypotheses being as in the theorem, if two cuppings $E \times F \to H$ coincide in dimension 0, then they coincide in dimension 1.*

Proof. For all $\alpha \in E^1(A)$ there exists $\xi \in E^0(X_A)$ such that $\alpha = \delta\xi$, and by hypothesis we have for $\beta \in F(B)$,

$$\alpha\beta = (\delta\xi)\beta = \delta(\xi\beta),$$

whence the corollary follows.

Theorem 1.4. *Let G be a group and let H be the cohomological functor H_G from Chapter I, such that $H^0(A) = A^G$ for A in $\mathrm{Mod}(G)$. Then a cupping*

$$H \times H \to H$$

such that in dimension 0, the cupping is induced by the bilinear map

$$(a, b) \mapsto \theta(a, b) \quad \text{for} \quad \theta : A \times \to C \quad \text{and} \quad a \in A^G, b \in B^G,$$

is uniquely determined by this condition.

Proof. This is just a special case of Theorem 1.2, since in Chapter I we proved the existence of the erasing functor (M, ε) necessary to prove uniqueness.

We shall always consider the functor H_G as having its structure of cup functor as we have just defined it in Theorem 1.4. Its existence will be proved in the next section.

If G is finite, then in the category $\mathrm{Mod}(G)$ we have an erasing functor $\mathbf{M}_G$ for the special cohomology functor $\mathbf{H}_G$. So we now formulate the general situation.

Let $E = (E^p), F = (F^q)$, and $H = (H^r)$ be three exact δ-functors on a multilinear category $\mathfrak{A}$. We suppose given a cupping $E \times F \to H$. As in Chapter I, suppose given an erasing functor $\mathbf{M}$ for E in dimensions $> p_0$. We say that $\mathbf{M}$ is **special** if for each bilinear map

$$\theta : A \times B \to C$$

in $\mathfrak{A}$, there exists bilinear maps

$$\mathbf{M}(\theta) : \mathbf{M}_A \times B \to \mathbf{M}_C \quad \text{and} \quad X(\theta) : X_A \times B \to X_C$$

such that the following diagram is commutative:

$$
\begin{array}{ccccc}
A \times B & \longrightarrow & \mathbf{M}_A \times B & \longrightarrow & X_A \times B \\
\theta \downarrow & & M(\theta) \downarrow & & M(\theta) \downarrow \\
C & \longrightarrow & \mathbf{M}_C & \longrightarrow & X_C
\end{array}
$$

Theorem 1.5 (right). *Let $E = (E^p), F = (F^q)$ and $H = (H^r)$ be three exact δ-functors on a multilinear category $\mathfrak{A}$. Suppose given a cupping $E \times F \to H$. Let $\mathbf{M}$ be a special erasing functor for E and for H in all dimensions. Then there is a commutative diagram associated with each bilinear map $\theta : A \times B \to C$:*

$$
\begin{array}{ccc}
E^p(X_A) \times F^q(B) & \longrightarrow & H^{p+q}(C) \\
\delta \downarrow & \mathrm{id} \downarrow & \delta \downarrow \\
E^{p+1}(A) \times F^q(B) & \longrightarrow & H^{p+1+1}(C)
\end{array}
$$

and the vertical maps are isomorphisms.

Proof. Clear.

Of course, we have the dual situation when $\mathbf{M}$ is a coerasing functor for E in dimensions $< p_0$. In this case, we say that $\mathbf{M}$ is **special** if for each θ there exist bilinear maps $\mathbf{M}(\theta)$ and $Y(\theta)$ making the following diagram commutative:

$$
\begin{array}{ccccc}
Y_A \times B & \longrightarrow & \mathbf{M}_A \times B & \longrightarrow & A \times B \\
{\scriptstyle Y(\theta)}\downarrow & & {\scriptstyle \mathbf{M}(\theta)}\downarrow & & {\scriptstyle \theta}\downarrow \\
Y_C & \longrightarrow & \mathbf{M}_C & \longrightarrow & C
\end{array}
$$

One then has:

Theorem 1.5 (left). *Let E, F, H be three exact δ-functors on a multilinear category $\mathfrak{A}$. Suppose given a cupping $E \times F \to H$. Let $\mathbf{M}$ be a special coerasing functor for E and H in all dimensions. Then for each bilinear map $\theta : A \times B \to C$ we have a commutative diagram*

$$
\begin{array}{ccccc}
E^p(A) & \times & F^q(B) & \longrightarrow & H^{p+q}(C) \\
{\scriptstyle \delta}\downarrow & & {\scriptstyle \mathrm{id}}\downarrow & & {\scriptstyle \mathrm{id}}\downarrow \\
E^{p+q+1}(Y_A)\times & & F^q(B) & \longrightarrow & H^{p+q+1}(C)
\end{array}
$$

and the vertical maps are isomorphisms.

Corollary 1.6. *Let E, F, H be as in the preceding theorems, with values $p = 0, 1$ (resp. $0, -1$); $q = 0$; $r = 0, 1$ (resp. $0, -1$). Let M be an erasing functor (resp. coerasing) for E and H. Suppose given two cuppings $E \times F \to H$ which coincide in dimension 0. Then they coincide in dimension 1 (resp. -1).*

We observe that the choice of indices $0, 1, -1$ is arbitrary, and the corollary applies mutatis mutandis to $p, p+1$, or $p, p-1$ for E, q arbitrary for F, and $p+q, p+q+1$ (resp. $p+q, p+q-1$) for H.

Corollary 1.7. *Let H be a cohomological functor on a multilinear category $\mathfrak{A}$. Suppose there exists a special erasing and coerasing functor on H. Suppose there are two cup functor structures on H, coinciding in dimension 0. Then these cuppings coincide in all dimensions.*

Proposition 1.8. *Let G be a finite group and G' a subgroup. Let $\mathbf{H}$ be the cohomological functor on $\mathrm{Mod}(G)$ such that $\mathbf{H}(A) = \mathbf{H}(G', A)$. Suppose given in addition an additional structure of cup functor on $\mathbf{H}$. Then the erasing functor $A \mapsto \mathbf{M}_G(A)$ (and the similar coerasing functor) are special.*

Proof. Let $\theta : A \times B \to C$ be bilinear. Viewing $\mathbf{M}_G$ as coerasing, we have the commutative diagram

$$
\begin{array}{ccccc}
I_G \otimes A \otimes B & \longrightarrow & \mathbf{Z}[G] \otimes A \otimes B & \longrightarrow & \mathbf{Z} \otimes A \otimes B \\
\downarrow & & \downarrow & & \downarrow \\
I_G \otimes C & \longrightarrow & \mathbf{Z}[G] \otimes C & \longrightarrow & \mathbf{Z} \otimes C
\end{array}
$$

where the vertical maps are defined by $\lambda \otimes a \otimes b \mapsto \lambda \otimes ab$. We have a similar diagram to the right for the erasing functor.

From the general theorems, we then obtain:

Theorem 1.9. *Let G be a group and $H = H_G$ the ordinary cohomological functor on $\mathrm{Mod}(G)$. Then for each multilinear map $A_1 \times \cdots \times A_n \to B$ in $\mathrm{Mod}(G)$, if we define*

$$\varkappa(a_1) \cdots \varkappa(a_n) = \varkappa(a_1 \cdots a_n) \quad \text{for} \quad a_i \in A_i^G,$$

then we obtain a multilinear functor, and a cupping

$$H^0 \times \cdots \times H^0 \to H^0.$$

A cup functor structure on H which is the above in dimension 0 is uniquely determined. The similar assertion holds when G is finite and H is replaced by the special functor $\mathbf{H} = \mathbf{H}_G$.

As mentioned previously, existence will be proved in the next section. For the rest of this section, we let H denote the ordinary or special cohomology functor on $\mathrm{Mod}(G)$, depending on whether G is arbitrary or finite.

Corollary 1.10. *Let G be a group and H the ordinary or special functor if G is finite. Let $A_1, A_2, A_3, A_{12}, A_{123}$ be G-modules. Suppose given multilinear maps in $\mathrm{Mod}(G)$:*

$$A_1 \times A_2 \to A_{12} \qquad\qquad A_{12} \times A_3 \to A_{123}$$

whose composite gives rise to a multilinear map

$$A_1 \times A_2 \times A_3 \to A_{123}.$$

Let $\alpha_i \in H^{p_i}(A_i)$. Then we have associativity,

$$(\alpha_1 \alpha_2)\alpha_3 = \alpha_1(\alpha_2 \alpha_3),$$

these cup products being taken relative to the multilinear maps as above.

Proof. One first reduces the theorem to the case when $A_{12} = A_1 \otimes A_2$ and $A_{123} = A_1 \otimes A_2 \otimes A_3$. The products on the right and on the left of the equation satisfy the axioms of a cup product, so we can apply the uniqueness theorem.

Remark. More generally, to define a cup functor structure on a cohomological functor H on an abelian category of abelian groups $\mathfrak{A}$, closed under tensor products, it suffices to give a cupping for two factors, i.e. $H \times H \to H$. Once this is done, let

$$\alpha_1 \in H^{r_1}(A_1), \dots, \alpha_n \in H^{r_n}(A_n).$$

We may then define

$$\alpha_1 \dots \alpha_n = (\dots (\alpha_1 \alpha_2)\alpha_3) \dots \alpha_n),$$

and one sees that this gives a structure of cup functor on H, using the universal property of the tensor product.

Corollary 1.11. *Let G be a group and $H = H_G$ the ordinary cup functor, or the special one if G is finite. Let $\theta : A \times B \to C$ be bilinear in $\mathrm{Mod}(G)$. Let $a \in A^G$, and let*

$$\theta_a : B \to C \quad \text{be defined by} \quad \theta_a(b) = ab.$$

Then θ_a is a morphism in $\mathrm{Mod}(G)$. If

$$H^q(\theta_a) = \theta_{a*} : H^q(B) \to H^q(C)$$

is the induced homomorphism, then

$$\varkappa(a)\beta = \theta_{a*}(\beta) \quad \text{for} \quad \beta \in H^q(B).$$

Proof. The first assertion is clear from the fact that $a \in A^G$ implies $\sigma(ab) = a\sigma b$. If $q = 0$ the second assertion amounts to the definition of the induced mapping. For the other values of q, we apply the uniqueness theorem to the cuppings $H^0 \times H \to H$ given either by the cup product or by the induced homomorphism, to conclude the proof.

Corollary 1.12. *Suppose G finite and $\theta : A \times B \to C$ bilinear in* $\mathrm{Mod}(G)$. *Then for $a \in A^G$ and $b \in B_{\mathbf{S}_G}$ we have*

$$\varkappa(a) \cup \maltese(b) = \maltese(ab).$$

Proof. A direct verification shows that the above formula defines a cupping of $\mathbf{H}^0$ and $(\mathbf{H}^0, \mathbf{H}^{-1})$ into $(\mathbf{H}^0, \mathbf{H}^{-1})$. Under the stated hypotheses, we clearly have $ab \in C_{\mathbf{S}_G}$, so the formula makes sense, and is valid.

§2. Existence

We shall see that on $\mathrm{Mod}(G)$ the cup product is induced by a product of cochains, and we shall give the explicit formula for cochains in the standard complex. First we give an axiomatic version so the reader sees the general situation in the framework of abelian categories. But as usual, the reader may think of $\mathrm{Mod}(G)$ and abelian groups for the categories mentioned in the next theorem.

Theorem 2.1. *Let $\mathfrak{A}$ be a multilinear category of abelian groups. Let*

$$A \mapsto Y(A)$$

be an exact functor of $\mathfrak{A}$ into the category of complexes in $\mathfrak{A}$. Suppose that for each bilinear map $\theta : A \times B \to C$ in $\mathfrak{A}$ we are given a bilinear map

$$Y^r(A) \times Y^s(B) \to Y^{r+s}(C)$$

which is functorial in A, B, C (covariant) and such that if $f \in Y^r(A)$ and $g \in Y^s(B)$, then

$$\delta(fg) = (\delta f)g + (-1)^r f(\delta g).$$

Let H be the cohomological functor associated to the functor Y. Then there exists on H a structure of cup functor, induced by the above bilinear map. This structure satisfies the property of the three exact sequences, as it is described below.

The **property of the three exact sequences** is the following. Consider three exact sequences in $\mathfrak{A}$:

$$0 \longrightarrow A' \longrightarrow A \longrightarrow A'' \longrightarrow 0$$

$$0 \longrightarrow B' \longrightarrow B \longrightarrow B'' \longrightarrow 0$$

$$0 \longrightarrow C' \longrightarrow C \longrightarrow C'' \longrightarrow 0$$

Suppose given a bilinear map $A \times B \to C$ in $\mathfrak{A}$ such that

$$A'B = 0, \qquad AB' \subset C', \qquad A'B \subset C', \qquad A''B'' \subset C''.$$

Then for $\alpha'' \in H^r(A'')$ and $\beta'' \in H^s(B'')$ we have

$$\delta(\alpha'' \cup \beta'') = (\delta\alpha'') \cup \beta'' + (-1)^r \alpha'' \cup (\delta\beta'').$$

It is a pain to write down in complete detail what amounts to a routine proof. One has to combine the additive construction with multiplicative considerations on the product of two complexes, cf. for instance Exercise 29 of *Algebra*, Chapter XX. More generally, we observe the following general fact. Let $\mathfrak{A}$ be a multilinear abelian category and let $\mathrm{Com}(\mathfrak{A})$ be the abelian category of complexes in $\mathfrak{A}$. Then we can make $\mathrm{Com}(\mathfrak{A})$ into a multilinear category as follows. Let K, L, M be three complexzes in $\mathfrak{A}$. We **define a bilinear map**

$$\theta : K \times L \to M$$

to be a family of bilinear maps

$$\theta_{r,s} : K^r \times L^s \to M^{r+s}$$

satisfying the condition

$$\delta_M(\theta_{r,s}(x,y)) = \theta_{r+1,s}(\delta_K x, y) + (-1)^r \theta_{r,s+1}(x, \delta_L y)$$

for $x \in K^r$ and $y \in L^s$. We have indexed the coboundaries $\delta_M, \delta_L, \delta_K$ according to the complexes to which they belong. If we omit all indices to simplify the notation, then the above condition reads

$$\delta(x \cdot y) = \delta x \cdot y + (-1)^r x \cdot \delta y.$$

Exercise 29 loc. cit. reproduces this formula for the universal bilinear map given by the tensor product. Then the cup product is induced by a product on representing cochains in the cochain complex. We leave the details to the reader. As we shall see, in practice, one can give explicit concrete formulas which allow direct verification.

We shall now make the theorem explicit for the standard complex and $\mathrm{Mod}(G)$.

Lemma 2.2. *Let G be a group and $Y(G, A)$ the homogeneous standard complex for $A \in \mathrm{Mod}(G)$. Let $A \times B \to C$ be bilinear in $\mathrm{Mod}(G)$. For $f \in Y^r(G, A)$ and $g \in Y^s(G, A)$ define the product fg by the formula*

$$(fg)(\sigma_0, \ldots, \sigma_{r+1}) = f(\sigma_0, \ldots, \sigma_r)g(\sigma_r, \ldots, \sigma_{r+s}).$$

Then this product satisfies the relation

$$\delta(fg) = (\delta f)g + (-1)^r f(\delta g).$$

Proof. Straightforward.

In terms of non-homogeneous cochains f, g in the non-homogeneous standard complex, the formula for the product is given by

$$(fg)(\sigma_1, \ldots, \sigma_{r+s}) = f(\sigma_1, \ldots, \sigma_r)(\sigma_1, \ldots, \sigma_r g(\sigma_{r+1}, \ldots, \sigma_{r+s})).$$

Theorem 2.3. *Let G be a group and let H_G be the ordinary cohomological functor on $\mathrm{Mod}(G)$. Then the product defined in Lemma 2.2 induces on H_G a structure of cup functor which satisfies the property of the three exact sequences. Furthermore in dimension 0, we have*

$$\varkappa(ab) = \varkappa(a)\varkappa(b) \qquad for\ a \in A^G\ and\ b \in B^G.$$

Proof. This is simply a special case of Theorem 2.1, taking the lemma into account, giving the explicit expression for the cupping in terms of cochains in the standard complex. The property of the three exact sequences is immediate from the definition of the coboundary. Indeed, if we are given cochains f'' and g'' representing α'' and β'', their coboundaries are defined by taking cochains f, g in A, B respectively mapping on f'', g'', so that fy maps on $f''g''$. The formula of the lemma implies the formula in the property of the three exact sequences.

We have the analogous result for finite groups and the special functor.

Lemma 2.4. *Let G be a finite group. Let X be a complete, $\mathbf{Z}[G]$-free, acyclic resolution of $\mathbf{Z}$, with augmentation ε. Let d be the boundary operation in X and let*

$$d' = d \otimes \mathrm{id}_X \qquad and \qquad d'' = \mathrm{id}_X \otimes d$$

be those induced in $X \otimes X$. Then there is a family of G-morphisms

$$h_{r,s} : X_{r+s} \longrightarrow X_r \otimes X_s \qquad for \quad -\infty < r, s < \infty,$$

satisfying the following conditions:

(i) $h_{r,s}d = d'h_{r+1,s} + d''h_{r,s+1}$

(ii) $(\varepsilon \otimes \varepsilon)h_{0,0} = \varepsilon$.

Proof. Readers will find a proof in Cartan-Eilenberg [CaE].

Theorem 2.5. *Let G be a finite group and $\mathbf{H}_G$ the special cohomology functor. Then there exists a unique structure of cup functor on $\mathbf{H}_G$ such that if $\theta : A \times B \to C$ is bilinear in $\mathrm{Mod}(G)$ then*

$$\varkappa(a)\varkappa(b) = \varkappa(ab) \quad for \quad a \in A^G, b \in B^G.$$

Furthermore, this cupping satisfies the three exact sequences property.

Proof. Let $A \in \mathrm{Mod}(G)$, and let X be the standard complex. Let $Y(A)$ be the cochain complex $\mathrm{Hom}_G(X, A)$. For

$$f \in Y^r(A) = \mathrm{Hom}_G(X^r, A) \qquad and \qquad g \in Y^s(A),$$

we can define the product $fg \in Y^{r+s}(A)$ by the composition of canonical maps

$$X_{r+s} \xrightarrow{h_{r,s}} X_r \otimes X_s \xrightarrow{f \otimes g} A \otimes B \xrightarrow{\theta'} C,$$

where θ' is the morphism in $\mathrm{Mod}(G)$ induced by θ, that is

$$fg = \theta'(f \otimes g)h_{r,s}.$$

One then verifies without difficulty the formula

$$\delta(fg) = (\delta f)g + (-1)^r f(\delta g),$$

and the rest of the proof is as in Theorem 2.3.

§3. Relations with subgroups

We shall tabulate a list of commutativity relations for the cup product with a group and its subgroups.

First we note that every multilinear map θ in $\mathrm{Mod}(G)$ induces in a natural way a multilinear map θ' in $\mathrm{Mod}(G')$ for every subgroup G' of G. Set theoretically, it is just θ.

Theorem 3.1. *Let G be a group and $\theta : A \times B \to C$ bilinear in $\mathrm{Mod}(G)$. Let G' be a subgroup of G. Let H denote the ordinary cup functor, or the special cup functor if G is finite, except when we deal with inflation in which case H denotes only the ordinary functor. Let the restriction* res *be from G to G'. Then:*

(1) $\mathrm{res}(\alpha\beta) = (\mathrm{res}\,\alpha)(\mathrm{res}\,\beta)$ *for* $\alpha \in H^r(G, A)$ *and* $\beta \in H^s(G, B)$.

(2) $\mathrm{tr}((\mathrm{res}\,\alpha)\beta') = \alpha(\mathrm{tr}\,\beta')$ *for* $\alpha \in H^r(G, A)$ *and* $\beta' \in H^s(G', B)$,

the transfer being taken from G' to G. Similarly,

$$\mathrm{tr}(\alpha'(\mathrm{res}\,\beta)) = (\mathrm{tr}\,\alpha')\beta \text{ for } \alpha' \in H^r(G', A) \text{ and } \beta \in H^s(G, B).$$

(3) *Let G' be normal in G. Then θ induces a bilinear map*

$$A^{G'} \times B^{G'} \to C^{G'},$$

and for $\alpha \in H^r(G/G', A^{G'}), \beta \in H^s(G/G', B^{G'})$ *we have*

$$\inf(\alpha\beta) = (\inf \alpha)(\inf \beta).$$

Proof. The formulas are immediate in dimension 0, i.e. for $r = s = 0$. In each case, the expressions on the left and on the right of the stated equality define separately a cupping of a cohomological functor into another, coinciding in dimension 0, and satisfying the conditions of the uniqueness theorem. The equalities are therefore valid in all dimensions. For example, in (1) we have two cuppings of $H_G \times H_G \to H_G$, given by

$$(\alpha, \beta) \mapsto \mathrm{res}(\alpha\beta) \quad \text{and} \quad (\alpha, \beta) \mapsto (\mathrm{res}\,\alpha)(\mathrm{res}\,\beta).$$

In (3), we find first a cupping

$$H_{G/G'} \times H_{G/G'} \to H_G$$

to which we apply the uniqueness theorem on the right. We let the reader write out the details. For the inflation, we may write explicitly one of these cuppings:

$$H(G/G', A^{G'}) \times H(G/G', B^{G'}) \xrightarrow{\mathrm{cup}} H(G/G', C^{G'}) \to H(G, C^G).$$

§4. The triplets theorem

We shall formulate for cup products the analogue of the triplets theorem. We shall reduce the proof to the preceding situation.

Theorem 4.1. *Let G be a finite group and $\theta : A \times B \to C$ bilinear in $\mathrm{Mod}(G)$. Fix $\alpha \in \mathbf{H}^p(G, A)$ for some index p. For each subgroup G' of G let $\alpha' = \mathrm{res}^G_{G'}(\alpha)$ be the restriction in $\mathbf{H}^p(G', A)$. For each integer s denote by*

$$\alpha'_s : \mathbf{H}^s(G', B) \to \mathbf{H}^s(G', C)$$

the homomorphism $\beta' \mapsto \alpha'\beta'$. Suppose there exists an index r such that α'_{r+1} is surjective, α'_r is an isomorphism, and α'_{r+1} is injective, for all subgroups G' of G. Then α'_s is an isomorphism for all s.

Proof. Suppose first that $r = 0$. We then know by Corollary 1.11 that α'_s is the homomorphism $(\theta_a)_*$ induced by an element $a \in A^G$, where $\theta_a : B \to C$ is defined by $b \mapsto \theta(a, b) = ab$. The theorem is therefore true if $p = 0$ by the ordinary triplets theorem. We note that the induced homomorphism is compatible with the restriction from G to G'.

We then prove the theorem in general by ascending and descending induction on p. For example, let us give the details in the case of descending induction to the left. We have $E = F = H$. There exists $\xi \in \mathbf{H}^r(G, X_A)$ such that $\alpha = \delta\xi$, where X_A is the cokernel in the dimension shifting exact sequence as in Theorem 1.14 of Chapter II. The restriction being a morphism of functors, we have $\alpha'_s = \delta\xi'_s$ for all s. It is clear that α'_s is an isomorphism (resp. is injective, resp. surjective) if and only if ξ'_s is an isomorphism (resp. is injective, resp. surjective). Thus we have an inductive procedure to prove our assertion.

§5. The cohomology ring and duality

Let A be a ring and suppose that the group G acts on the additive group of A, i.e. that this additive group is in $\mathrm{Mod}(G)$. We say that A is a G-**ring** if in addition we have

$$\sigma(ab) = (\sigma a)(\sigma b) \quad \text{for all} \quad \sigma \in G, a, b \in A.$$

Suppose A is a G-ring. Then multiplication of n elements of A is a multilinear map in the multilinear category $\mathrm{Mod}(G)$.

Let us denote by $H(A)$ the direct sum

$$H(A) = \bigoplus_{-\infty}^{\infty} H^p(A),$$

where H is the ordinary functor on $\mathrm{Mod}(G)$, or special functor in case G is finite. Then $H(A)$ is a graded ring, multiplication being first defined for homogeneous elements $\alpha \in H^p(A)$ and $\beta \in H^q(A)$ by the cup product, and then on direct sums by linearity, that is

$$\left(\sum \alpha^{(p)} \right) \left(\sum \beta^{(q)} \right) = \sum_r \left(\sum_{p+q=r} \alpha^{(p)} \beta^{(q)} \right).$$

We then say that $H(A)$ is the **cohomology ring** of A.

One verifies at once that if A is a commutative ring, then $H(A)$ is anti-commutative, that is if $\alpha \in H^p(A)$ and $\beta \in H^q(A)$ then

$$\alpha\beta = (-1)^{pq}\beta\alpha.$$

Since by definition a ring has a unit element, we have $1 \in A^G$ and $\varkappa(1)$ is the unit element of $H(A)$. Indeed, for $\beta \in H^q(A)$ we have

$$\varkappa(1)\beta = \theta_{1*}\beta = \beta,$$

because $\theta_1 : a \mapsto 1a = a$ is the identity.

Let A be a G-ring and $B \in \mathrm{Mod}(G)$. Suppose that B is a left A-module, compatible with the action of G, that is the map

$$A \times B \to B$$

defined by the action of A on B is bilinear in the multilinear category $\mathrm{Mod}(G)$. We then obtain a product

$$H^p(A) \times H^q(B) \to H^{p+q}(B)$$

which we can extend by linearity so as to make the direct sum $H(B) = \bigoplus H^q(B)$ into a graded $H(A)$-module. The unit element of $H(A)$ acts as the identity on $H(B)$ according to the previous remarks.

Let $B, C \in \mathrm{Mod}(G)$. There is a natural map

$$\mathrm{Hom}(B, C) \times B \to C$$

defined by $(f, b) \mapsto f(b)$. This map is bilinear in the multilinear category $\mathrm{Mod}(G)$, because we have

$$([\sigma]f)(\sigma b) = \sigma f \sigma^{-1} \sigma b = \sigma(fb)$$

for $f \in \mathrm{Hom}(B, C)$ and $b \in B$. Thus we obtain a product

$$(\varphi, \beta) \mapsto \varphi\beta, \quad \text{for} \quad \varphi \in H^p(\mathrm{Hom}(B, C)) \quad \text{and} \quad \beta \in H^q(B).$$

Theorem 5.1. *Let*

$$0 \to B' \to B \to B'' \to 0$$

be a short exact sequence in $\mathrm{Mod}(G)$, *let* $C \in \mathrm{Mod}(G)$, *and suppose the* Hom *sequence*

$$0 \to \mathrm{Hom}(B'', C) \to \mathrm{Hom}(B, C) \to \mathrm{Hom}(B', C') \to 0$$

is exact. Then for $\beta'' \in H^{q-1}(B'')$ *and* $\varphi' \in H^p(\mathrm{Hom}(B', C))$, *we have*

$$(\delta\varphi')\beta'' + (-1)^p\varphi'(\delta\beta'') = 0.$$

Proof. We consider the three exact sequences:

$$0 \to \operatorname{Hom}(B'', C) \to \operatorname{Hom}(B, C) \to \operatorname{Hom}(B', C) \to 0$$

$$0 \to B' \to B \to B'' \to 0.$$

$$0 \to C \to C \to 0 \to 0,$$

and a bilinear map from $\operatorname{Mod}(G)$, in the middle, inducing mappings as in the existence theorem for the cup product. Since $\varphi' \beta'' = 0$, we find the present result as a special case.

We may rewrite the result of Theorem 5.1 in the form of a diagram

$$
\begin{array}{ccccc}
H^r(\operatorname{Hom}(B', C)) & \longrightarrow & H^s(B') & \longrightarrow & H^{r+s}(C) \\
\delta \downarrow & & \delta \downarrow & & \mathrm{id} \downarrow \\
H^{r+1}(\operatorname{Hom}(B'', C)) & \longrightarrow & H^{s-1}(B'') & \longrightarrow & H^{r+s}(C)
\end{array}
$$

which has character $(-1)^{r+1}$.

In dimension 0, we find:

Proposition 5.2. *Let* $f \in \operatorname{Hom}_G(B, C) = (\operatorname{Hom}(B, C))^G$, *and* $\beta \in H^r(B)$. *Then*

$$\varkappa(f)\beta = f_* \beta.$$

If G *is finite and* **H** *is the special functor, then*

$$\varkappa(f) \cup \varkappa(b) = \varkappa(f(b)) \quad \text{for} \quad b \in B_{\mathbf{S}_B}.$$

Proof. In dimension 0, this is an old result. The assertion concerning dimensions -1 and 0 is a special case of the uniqueness theorem, Corollary 1.12.

There is a homomorphism

$$h_{r,s} : H^r(\operatorname{Hom}(B, C)) \to \operatorname{Hom}(H^r(B), H^{r+s}(C)),$$

obtained from the bilinear map

$$H^r(\operatorname{Hom}(B, C)) \times H^s(B) \to H^{r+s}(C).$$

In important cases, we shall see that $h_{r,s}$ is an isomorphism. We shall now give a criterion for this in the case of the special functor.

Theorem 5.3. *Let G be finite, and $\mathbf{H} = \mathbf{H}_G$ the special functor. Suppose that for C fixed and B variable in $\operatorname{Mod}(G)$, and two fixed integers p_0, q_0 the map h_{p_0, q_0} is an isomorphism. Then $h_{p,q}$ is an isomorphism for all p, q such that $p + q = p_0 + q_0$.*

Proof. We are going to use dimension shifting. We consider the exact sequence

$$0 \to I \otimes B \to \mathbf{Z}[G] \otimes B \to \mathbf{Z} \otimes B = B \to 0,$$

which we hom into C. Since the sequence splits, we obtain an exact sequence

$$0 \to \operatorname{Hom}(B, C) \to \operatorname{Hom}(\mathbf{Z}[G] \otimes B, C) \to \operatorname{Hom}(I \otimes B, C) \to 0.$$

Applying the diagram following theorem 5.2, we find:

$$
\begin{array}{ccc}
\mathbf{H}^p(\operatorname{Hom}(I \otimes B, C)) & \xrightarrow{\ h_{p,q}\ } & \operatorname{Hom}(\mathbf{H}^q(I \otimes B), \mathbf{H}^{p+q}(C)) \\
{\scriptstyle \delta} \downarrow & & \downarrow {\scriptstyle (\delta, 1)} \\
\mathbf{H}^{p+1}(\operatorname{Hom}(B, C)) & \xrightarrow[\ h_{p+1, q+1}\ ]{} & \operatorname{Hom}(\mathbf{H}^{q-1}(B), \mathbf{H}^{p+q}(C)).
\end{array}
$$

The vertical coboundaries are isomorphisms because the middle object in the exact sequence is G-regular, and so annuls the cohomology. This concludes the proof going from p to $p + 1$. Going the other way, we use the other exact sequence

$$0 \to B \to \mathbf{Z}[G] \otimes B \to J \otimes B \to 0,$$

and we let the reader finish this side of the proof.

As an application, we shall prove a duality theorem. Let B be an abelian group. As before, we define its dual group by $\hat{B} = \operatorname{Hom}(B, \mathbf{Q}/\mathbf{Z})$. It is the group of characters of finite order, which we consider as a discrete group. Its elements will be called simply **characters**. Let $B \in \operatorname{Mod}(G)$. We consider B as an abelian group to get $\hat{B}$.

We have an isomorphism

$$\varkappa^{-1} : \mathbf{H}^{-1}(\mathbf{Q}/\mathbf{Z}) \to (\mathbf{Q}/\mathbf{Z})_n$$

between $\mathbf{H}^{-1}(\mathbf{Q}/\mathbf{Z})$ and the elements of order $n = (G : e)$ in $\mathbf{Q}/\mathbf{Z}$.

In addition, we have a bilinear map in $\mathrm{Mod}(G)$:

$$\hat{B} \times B \to \mathbf{Q}/\mathbf{Z},$$

and consequently a corresponding bilinear map of abelian groups

$$\mathbf{H}^{-q}(\hat{B}) \times \mathbf{H}^{q-1}(B) \to \mathbf{H}^{-1}(\mathbf{Q}/\mathbf{Z}).$$

Theorem 5.4. Duality Theorem. *The homomorphism*

$$\mathbf{H}^{-q}(\hat{B}) \to \mathbf{H}^{q-1}(B)^{\wedge}$$

which to each $\varphi \in \mathbf{H}^{-q}(B)$ *associates the character* $\beta \mapsto \mathrm{ж}^{-1}(\varphi\beta)$, *is an isomorphism, so we have*

$$\mathbf{H}^{-q}(\hat{B}) = \mathbf{H}^{q-1}(B)^{\wedge}.$$

Proof. From the definitions, we see that the theorem amounts to proving that $h_{-q,q-1}$ is an isomorphism. According to the preceding theorem, it suffices to show that $h_{0,-1}$ is an isomorphism. Since $h_{0,q}$ is an induced homomorphism, we can make $h_{0,-1}$ explicit in the present case, as follows. We have a homomoprhism

$$\hat{B}^G / \mathrm{S}_G \hat{B} \to (B_{\mathrm{S}_G} / IB)^{\wedge},$$

obtained by associating to an element $f \in \hat{B}$ the character $b \mapsto f(b)$ for $b \in B_{\mathrm{S}_G}$. We have to prove that this map is an isomorphism.

For the surjectivity, let $f_0 : B_{\mathrm{S}} \to (\mathbf{Q}/\mathbf{Z})_n$ be a homomorphism vanishing on IB. We can extend f_0 to a homomorphism f of B into $\mathbf{Q}/\mathbf{Z}$ because $\mathbf{Q}/\mathbf{Z}$ is injective. Furthermore f is in $\hat{B}^G$ because

$$f(\sigma b) - \sigma f(b) = f(\sigma b) - f(b) = f(\sigma b - b) = 0$$

by hypothesis. This proves surjectivity.

For the injectivity, let $f \in \hat{B}^G$ and suppose $f(B_{\mathrm{S}}) = 0$. Since B/B_{S} is isomorphic to $\mathrm{S}B$, there exists $g \in (\mathrm{S}B)^{\wedge}$ such that

$$f(b) = g(\mathrm{S}b) \quad \text{for} \quad b \in B.$$

We extend g to a homomorphism of B into $\mathbf{Q}/\mathbf{Z}$, denoted by the same letter. Then $f = \mathbf{S}g$, because

$$(\mathbf{S}g)(b) = \sum \sigma g \sigma^{-1} b = \sum g \sigma^{-1} b = g\left(\sum \sigma^{-1} b\right) = g\mathbf{S}b = f(b).$$

This concludes the proof of hte duality theorem.

Consider the special case when $B = \mathbf{Z}$. Then

$$\hat{B} = \hat{Z} = \operatorname{Hom}(\mathbf{Z}, \mathbf{Q}/\mathbf{Z})$$

and we find:

Corollary 5.5. $H^{-q}(\mathbf{Q}/\mathbf{Z}) \approx H^{q-1}(\mathbf{Z})^{\wedge}$.

Applying the coboundary homomorphism arising from the exact sequence

$$0 \to \mathbf{Z} \to \mathbf{Q} \to \mathbf{Q}/\mathbf{Z} \to 0,$$

we obtain:

Corollary 5.6. *The following diagram is commutative:*

$$
\begin{array}{ccc}
H^{-p-1}(\mathbf{Q}/\mathbf{Z}) \times H^p(\mathbf{Z}) & \longrightarrow & H^{-1}(\mathbf{Q}/\mathbf{Z}) \\
\downarrow{\scriptstyle \delta} \qquad\qquad \downarrow{\scriptstyle \mathrm{id}} & & \downarrow{\scriptstyle \delta} \\
H^{-p}(\mathbf{Z}) \quad \times \ H^p(\mathbf{Z}) & \longrightarrow & H^0(\mathbf{Z})
\end{array}
$$

The vertical maps are isomorphisms, and thus

$$H^{-p}(\mathbf{Z}) \approx H^p(\mathbf{Z})^{\wedge}.$$

Proof. Since $\mathbf{Q}$ is uniquely divisible by n, its cohomology groups are trivial, and hence the coboundaries are isomorphisms, so the corollary is clear.

Corollary 5.7. *Let $M \in \operatorname{Mod}(G)$ be $\mathbf{Z}$-free. Then one has a commutative diagram:*

$$
\begin{array}{ccc}
H^{p-1}(\operatorname{Hom}(M, Q/Z)) \times H^{-p}(M) & \longrightarrow & H^{-1}(\mathbf{Q}/\mathbf{Z}) \\
\downarrow{\scriptstyle \delta} \qquad\qquad\qquad \downarrow{\scriptstyle \mathrm{id}} & & \downarrow{\scriptstyle \delta} \\
H^p(\operatorname{Hom}(M, \mathbf{Z})) \quad \times \ H^{-p}(M) & \longrightarrow & H^0(\mathbf{Z})
\end{array}
$$

where the vertical maps are isomorphisms. Thus we obtain a canonical isomorphism

$$\mathbf{H}^p(\mathrm{Hom}(M, \mathbf{Z})) \approx \mathbf{H}^{-p}(M)^\wedge.$$

Proof. This is an immediate consequence of the fact that $\mathbf{Q}$ is G-regular (the identity is the trace of $1/n$), and so $\mathrm{Hom}(M, A)$ is also G-regular, and so annuls cohomology. The sequence

$$0 \to \mathrm{Hom}(M, \mathbf{Z}) \to \mathrm{Hom}(M, \mathbf{Q}) \to \mathrm{Hom}(M, \mathbf{Q}/\mathbf{Z}) \to 0$$

is exact. We can then apply the definition of the cup functor to conclude the proof.

It will be convenient to use the following terminology. Suppose given a bilinear map of finite abelian groups

$$F' \times F \to \mathbf{Q}/\mathbf{Z}.$$

This induces a homomorphism $F' \to F^\wedge$. If $F' \to F^\wedge$ is an isomorphism, then we say that F' and F are in **perfect duality** under the bilinear map. Each of Theorem 5.4, Corollary 5.5 and Corollary 5.6 establish a perfect duality of cohomology groups in their conclusions, when B is, say, finitely generated.

§6. Periodicity

In Chapter I, we saw that the cohomology of a finite cyclic group is periodic. We shall now give a general criterion for periodicity for an arbitrary finite group G. *This section will not be used in what follows.*

Let $r \in \mathbf{Z}$ be fixed. An element $\zeta \in \mathbf{H}^r(G, \mathbf{Z})$ will be said to be a **maximal generator** if ζ generates $\mathbf{H}^r(G, \mathbf{Z})$ and if ζ is of finite order $(G : e)$.

Theorem 6.1. *Let G be a finite group and $\zeta \in \mathbf{H}^r(G, \mathbf{Z})$. The following properties are equivalent.*

MAX 1. ζ *is a maximal generator.*

MAX 2. ζ *is of order $(G : e)$.*

MAX 3. *There exists an element $\zeta^{-1} \in \mathbf{H}^{-r}(G, \mathbf{Z})$*

such that $\zeta^{-1}\zeta = 1$.

MAX 4. For all $A \in \mathrm{Mod}(G)$, the map

$$\alpha \mapsto \zeta\alpha \quad \text{of} \quad \mathbf{H}^i(G, A) \to \mathbf{H}^{i+r}(G, A)$$

is an isomorphism for all i.

Proof. That **MAX 1** implies **MAX 2** is trivial.

Assume **MAX 2**. Suppose ζ has order $(G : e)$. Since $\mathbf{H}^{-r}(\mathbf{Z})$ is dual to $\mathbf{H}^r(\mathbf{Z})$ by Corollary 5.6, the existence of ζ^{-1} follows from the definition of the dual group, so **MAX 3** is satisfied.

Assume **MAX 3**. The maps

$$\alpha \mapsto \zeta\alpha \text{ for } \alpha \in H^i(A) \text{ and } \beta \mapsto \zeta^{-1}\beta \text{ for } \beta \in \mathbf{H}^{i+r}(A)$$

are inverse to each other, up to a power $(-1)^r$, and are therefore isomorphisms, thus proving **MAX 4**.

Assume **MAX 4**. We take $A = \mathbf{Z}$ and $i = 0$ in the preceding assertion, and we use the fact that $\mathbf{H}^0(\mathbf{Z})$ is cyclic of order $(G : e)$. Then **MAX 1** follows at once.

The uniqueness of ζ^{-1} in Theorem 6.1, satisfying condition **MAX 3**, is clear, taking into account that $\mathbf{H}^{-r}(\mathbf{Z})$ is the dual group of $\mathbf{H}^r(\mathbf{Z})$.

Proposition 6.2. *Let $\zeta \in \mathbf{H}^r(G, \mathbf{Z})$ be a maximal generator. Then so is ζ^{-1}. If ζ_1 is a maximal generator of $\mathbf{H}^s(G, \mathbf{Z})$ for some s, then $\zeta\zeta_1$ is a maximal generator.*

Proof. The first assertion follows from **MAX 3**; the second from **MAX 4**.

An integer m will be said to be a **cohomology period** of G if $\mathbf{H}^m(G, \mathbf{Z})$ contains a maximal generator, or in other words, $\mathbf{H}^m(G, \mathbf{Z})$ is cyclic of order $(G : e)$. The anticommutativity of the cup product shows that a period is even.

Proposition 6.3. *Suppose m is a cohomology period of G. Let U be a subgroup of G and let $\zeta \in \mathbf{H}^m(G, \mathbf{Z})$ be of order $(G : e)$.*

Then $\mathrm{res}_U^G(\zeta)$ *has order* $(U : e)$ *and* m *is a cohomology period of* U.

Proof. Since

$$\mathrm{tr}_G^U \mathrm{res}_U^G(\zeta) = (G : U)\zeta,$$

it follows that the order of the restriction of ζ to U is at least $(U : e)$. Since it is at most equal to $(U : e)$, it is a period.

Proposition 6.4. *Let* G_p *be a* p-*Sylow subgroup of* G *and let* $\zeta \in \mathrm{H}^r(G_p, \mathbf{Z})$ *be a maximal generator. Let* n *be a positive integer such that*

$$k^n \equiv 1 \mod (G_p : e)$$

for all integers k *prime to* p. *Then* $\zeta^n \in \mathrm{H}^{nr}(G_p, \mathbf{Z})$ *is stable, i.e.* $\sigma_*(\zeta^n) = \zeta^n$ *for all* $\sigma \in G$, *and*

$$\mathrm{tr}_G^{G_p}(\zeta^n)$$

has order $(G_p : e)$.

Proof. Since σ_* is an isomorphism, and since the restriction of a maximal generator is a maximal generator, one concludes that the elements

$$\beta = \mathrm{res}_{G_p}^{G_p \cap G_p[\sigma]}(\zeta^n) \quad \text{and} \quad \mathrm{res}_{G_p \cap G_p[\sigma]}^{G_p[\sigma]} \circ \sigma_*(\zeta^n) = \lambda$$

are both maximal generators in $\mathrm{H}^r(G_p \cap G_p[\sigma], \mathbf{Z})$. Hence there exists an integer k prime to p such that one is equal to k times the other, i.e. $k\beta = \lambda$. Taking the n-th power we get

$$k^n \beta^n = \lambda^n.$$

From the definition of n, with the fact that $(G_p : e)$ kills β and λ, and with the commutativity of the cup product and the indicated operations, we find that ζ^n is stable. This being the case, we know from Proposition 1.10 of Chapter II that

$$\mathrm{res}_{G_p}^G \circ \mathrm{tr}_G^{G_p}(\zeta^n) = (G : G_p)\zeta^n.$$

Since $(G : G_p)$ is prime to p, it follows that the transfer followed by the restriction is injective on the subgroup generated by ζ^n. Thus the transfer is injective on this subgroup. From this one sees that the period of this transfer is the same as that of ζ^n, whence the same as that of ζ, thus concluding the proof.

Corollary 6.5. *Let G be a finite group. Then G admits a cohomological period > 0 if and only if each Sylow subgroup G_p has a period > 0.*

Proof. If G has a period > 0, the proposition shows that G_p also has one. Conversely, suppose that $\zeta'_p \in \mathbf{H}^r(G_p, \mathbf{Z})$ is a maximal generator. Let

$$\zeta_p = \mathrm{tr}_G^{G_p}(\zeta'_p).$$

Then the order of ζ_p is the same as that of ζ'_p by Proposition 6.4. Let

$$\zeta = \sum \zeta_p.$$

Then ζ is an element of order $(G : e)$, so G has a cohomological period > 0, as was to be shown.

The preceding corollary reduces the study of periodicity to p-groups. One can show easily:

Proposition 6.6. *Let $G = G_p$ be a p-group. Then G admits a cohomological period > 0 if and only if G is cyclic or G is a generalized quaternion group.*

We omit the proof.

§7. The theorems of Tate-Nakayama

We shall now go back to the theorem concerning the splitting module for a class module as in Chapter III, §3. We recall that if $A' \in \mathrm{Mod}(G)$ is cohomologically trivial and M is a G-module without torsion, then $A' \otimes M$ is cohomologically trivial.

Theorem 7.1. (Tate). *Let G be a finite group, $M \in \mathrm{Mod}(G)$ without torsion, $A \in \mathrm{Mod}(G)$ a class module, and $\alpha \in \mathbf{H}^2(G, A)$ fundamental. Let*

$$\alpha_r : \mathbf{H}^r(G, M) \to \mathbf{H}^{r+2}(G, A \otimes M)$$

be the cup product relative to the bilinear map $A \times M \to A \otimes M$, i.e. such that

$$\alpha_r(\lambda) = \alpha \cup \lambda \quad for \quad \lambda \in \mathbf{H}^r(G, M).$$

Then α_r is an isomorphism for all $r \in \mathbf{Z}$.

Proof. As in the main theorem on cohomological triviality (Theorem 3.1 of Chapter III, we have exact sequences

$$0 \longrightarrow A \longrightarrow A' \longrightarrow I \longrightarrow 0$$

$$0 \longrightarrow A \otimes M \longrightarrow A' \otimes M \longrightarrow I \otimes M \longrightarrow 0.$$

The exactness of the second sequence is due to the fact that the first one splits.

In addition, $A' \otimes M$ is cohomologically trivial. Let us put $\beta = \delta\zeta$. Then

$$\alpha_r(\lambda) = \alpha\lambda = (\delta\beta)\lambda = \delta(\beta\lambda).$$

If we now use the exact sequences

$$0 \longrightarrow I \longrightarrow \mathbf{Z}[G] \longrightarrow \mathbf{Z} \longrightarrow 0$$

$$0 \longrightarrow I \otimes M \longrightarrow \mathbf{Z}[G] \otimes M \longrightarrow \mathbf{Z} \otimes M \longrightarrow 0,$$

we find

$$\alpha_r(\lambda) = \delta\delta(\zeta\lambda).$$

The coboundaries δ are isomorphisms, in one case because $\mathbf{Z}[G] \otimes M$ is G-regular, in the other case by the main theorem on cohomological triviality. To show that α_r is an isomorphism, it will suffice to show that $\zeta_r : \lambda \mapsto \zeta\lambda$ is an isomorphism. But this is clear because it is the identity, as one sees by making explicit the canonical isomorphism $\mathbf{Z} \otimes M \to M$. This concludes the proof of Theorem 7.1.

We can rewrite the commutative diagram arising from the theorem in the following manner.

$$
\begin{array}{ccc}
\mathbf{H}^0(Z) \times \mathbf{H}^r(M) & \longrightarrow & \mathbf{H}^r(Z \otimes M) = \mathbf{H}^r(M) \\
\delta \downarrow \quad \text{id} \downarrow & & \delta \downarrow \\
\mathbf{H}^1(I) \times \mathbf{H}^r(M) & \longrightarrow & \mathbf{H}^{r+1}(I \otimes M) \\
\delta \downarrow \quad \text{id} \downarrow & & \delta \downarrow \\
\mathbf{H}^2(A) \times \mathbf{H}^r(M) & \longrightarrow & \mathbf{H}^{r+2}(A \otimes M).
\end{array}
$$

The vertical maps δ are isomorphisms, and the cup product on top corresponds to the bilinear map $\mathbf{Z} \times M \to \mathbf{Z} \otimes M = M$, so the isomorphism induced by ζ_r is the identity.

If we take $M = \mathbf{Z}$ and $r = -2$, we find

$$\mathbf{H}^2(A) \times \mathbf{H}^{-2}(\mathbf{Z}) \to \mathbf{H}^0(A).$$

We know that $\mathbf{H}^{-2}(\mathbf{Z}) = G/G^c$ and so we find an isomorphism

$$G/G^c \approx \mathbf{H}^0(A) = A^G/S_G A.$$

We shall make this isomorphism more explicit below.

We also obtain an analogous theorem by taking Hom instead of the tensor product, and by using the duality theorem.

Theorem 7.2. *Let G be a finite group, $M \in \mathrm{Mod}(G)$ and $\mathbf{Z}$-free, $A \in \mathrm{Mod}(G)$ a class module. Then for all $r \in \mathbf{Z}$, the bilinear map of the cup product*

$$\mathbf{H}^r(G, \mathrm{Hom}(M, A)) \times \mathbf{H}^{2-r}(G, M) \to \mathbf{H}^2(G, A)$$

induces an isomorphism

$$\mathbf{H}^r(G, \mathrm{Hom}(M, A)) \approx \mathbf{H}^{2-r}(G, M)^{\wedge}.$$

Proof. We shift dimensions on A twice. Since A' and $\mathbf{Z}[G]$ are $\mathbf{Z}$-free, it follows that the sequences

$$0 \longrightarrow \mathrm{Hom}(M, \mathbf{Z}) \longrightarrow \mathrm{Hom}(M, A') \longrightarrow \mathrm{Hom}(M, I) \longrightarrow 0$$

$$0 \longrightarrow \mathrm{Hom}(M, I) \longrightarrow \mathrm{Hom}(M, \mathbf{Z}[G]) \longrightarrow \mathrm{Hom}(M, \mathbf{Z}) \longrightarrow 0$$

are exact. By the definition of the cup product one finds commutative diagrams as follows, where the vertical maps are isomorphisms.

$$
\begin{array}{ccccc}
\mathbf{H}^{r-2}(\mathrm{Hom}(M, \mathbf{Z})) & \times & \mathbf{H}^{2-r}(M) & \longrightarrow & \mathbf{H}^0(\mathbf{Z}) \\
\downarrow{\scriptstyle \delta} & & \downarrow{\scriptstyle \mathrm{id}} & & \downarrow{\scriptstyle \delta} \\
\mathbf{H}^{r-1}(\mathrm{Hom}(M, I)) & \times & \mathbf{H}^{2-r}(M) & \longrightarrow & \mathbf{H}^1(I) \\
\downarrow{\scriptstyle \delta} & & \downarrow{\scriptstyle \mathrm{id}} & & \downarrow{\scriptstyle \delta} \\
\mathbf{H}^r(\mathrm{Hom}(M, A)) & \times & \mathbf{H}^{2-r}(M) & \longrightarrow & \mathbf{H}^2(A).
\end{array}
$$

The bilinear map on top is that of Corollary 5.7, and the theorem follows.

Selecting $M = \mathbf{Z}$ we get for $r = 0$:

$$\mathbf{H}^0(A) \times \mathbf{H}^2(\mathbf{Z}) \to \mathbf{H}^2(A),$$

this being compatible with the bilinear map

$$A \otimes \mathbf{Z} \to A \quad \text{such that} \quad (a, n) \mapsto na.$$

We know that $\delta : \mathbf{H}^1(\mathbf{Q/Z}) \to \mathbf{H}^2(\mathbf{Z})$ is an isomorphism, so we find the pairing

$$\mathbf{H}^0(G, A) \times \mathbf{H}^1(\mathbf{Q/Z}) = \hat{G} \to \mathbf{H}^2(G, A)$$

which is a perfect duality since G is finite. One can give an explicit determination of this duality in terms of standard cocycles as follows.

Theorem 7.3. *Let A be a class module in* $\mathrm{Mod}(G)$. *Then the perfect duality between $H^0(A)$ and $\hat{G}$ is induced by the following pairing. For $a \in A^G$ and a character $\chi : G \to \mathbf{Q/Z}$, we get a 2-cocycle*

$$a_{\sigma,\tau} = [\chi'(\sigma) + \chi'(\tau) - \chi'(\sigma\tau)]a,$$

where χ' is a lifting of χ in $\mathbf{Q}$. The expression in brackets is a 2-cocycle of G in $\mathbf{Z}$. The cocycle $(a_{\sigma,\tau})$ represents the class $\varkappa(a) \cup \delta\chi$.

Thus the perfect duality arises from a bilinear map

$$A^G \times \hat{G} \to \mathbf{H}^2(G, A)$$

which we may write

$$(a, \chi) \mapsto a \cup \delta\chi,$$

whose kernel on the left is $S_G A$, and the kernel on the right is 0.

§8. Explicit Nakayama maps

Throughout this section we let G be a finite group.

In Chapter I, §4 we had an isomorphism

$$\mathbf{H}^{-1}(G, \mathbf{Z}) \xrightarrow{\approx} G/G^c$$

by means of a sequence of isomorphisms

$$\mathbf{H}^{-2}(\mathbf{Z}) \approx \mathbf{H}^{-1}(I) \approx I/I^2 \approx G/G^c.$$

If $\tau \in G$, we denote by ζ_τ the element of $\mathbf{H}^{-2}(\mathbf{Z})$ corresponding to the coset τG^c in G/G^c. So by definition

$$\zeta_\tau = \delta^{-1}(\text{\ding{95}}(\tau - e)),$$

where δ is the coboundary associated to the exact sequence

$$0 \to I_G \to \mathbf{Z}[G] \to \mathbf{Z} \to 0.$$

On the other hand, we now have a cup product

$$\mathbf{H}^r(A) \times \mathbf{H}^{-2}(\mathbf{Z}) \to \mathbf{H}^{r-2}(A)$$

for $A \in \mathrm{Mod}(G)$, associated to the natural bilinear map $\mathbf{Z} \times A \to A$. We are going to make this cup product explicit for $r \geq 1$, in terms of cochains from the standard complex, and the description of $\mathbf{H}^{-2}(\mathbf{Z})$ given above.

To start, we give a special case of the cup product under dimension shifting. We consider as usual the exact sequence

$$0 \to I \to \mathbf{Z}[G] \to \mathbf{Z} \to 0$$

and its dual

$$0 \to \mathrm{Hom}(\mathbf{Z}, A) \to \mathrm{Hom}(\mathbf{Z}[G], A) \to \mathrm{Hom}(I, A) \to 0.$$

We have a bilinear map in $\mathrm{Mod}(G)$

$$\mathrm{Hom}(\mathbf{Z}[G], A) \times \mathbf{Z}[G] \to A \quad \text{given by} \quad (f, \lambda) \mapsto f(\lambda).$$

There results a pairing of these two exact sequences into the exact sequence
$$0 \to A \to A \to 0 \to 0,$$
to which one can apply the commutative diagram following Theorem 5.1 to get:

Proposition 8.1. *The following diagram has character -1:*

$$
\begin{array}{ccccc}
\mathbf{H}^0(\mathrm{Hom}(I,A)) & \times & \mathbf{H}^{-1}(I) & \longrightarrow & \mathbf{H}^{-1}(A) \\
\Big\downarrow \delta & & \Big\downarrow \delta & & \Big\downarrow \mathrm{id} \\
\mathbf{H}^1(\mathrm{Hom}(\mathbf{Z},A)) & \times & \mathbf{H}^{-2}(\mathbf{Z}) & \longrightarrow & \mathbf{H}^{-1}(A)
\end{array}
$$

On the other hand, we know that in dimension 0, the cup product is given by the induced morphisms. By Corollary 1.12, we see that the cup product in the top line is given by the maps

$$\varkappa(f) \cup \varkappa(\sigma - e) = \varkappa(f(\sigma - e)) \quad \text{for} \quad f \in \mathrm{Hom}_G(I,A).$$

We now pass to the general case.

Theorem 8.2. *Let* $a = a(\sigma_1, \ldots, \sigma_r)$ *be a standard cochain in* $C^r(G,A)$ *for* $r \geq 1$. *For each* $\tau \in G$, *define a map*

$$a \mapsto a * \tau \quad \text{of} \quad C^r(G,A) \to C^{r-2}(G,A)$$

by the formulas:

$$
\begin{aligned}
(a * \tau)(\cdot) &= a(\tau) && \text{if } r = 1 \\
(a * \tau)(\cdot) &= \sum_{\rho \in G} a(\rho, \tau) && \text{if } r = 2 \\
(a * \tau)(\sigma_1, \ldots, \sigma_{r-2}) &= \sum_{\rho \in G} a(\sigma_1, \ldots, \sigma_{r-2}, \rho, \tau) && \text{if } r > 2.
\end{aligned}
$$

Then for $r \geq 1$ *we have the relation*

$$(\delta a) * \tau = \delta(a * \tau).$$

If a *is a cocycle representing an element* α *of* $\mathbf{H}^r(G,A)$, *then* $(a * \tau)$ *represents* $\alpha \cup \zeta_\tau \in \mathbf{H}^{r-2}(G,A)$.

Proof. Let us first give the proof for $r = 1$ and 2. We let $a = a(\sigma)$ be a 1-cochain. We find

$$((\delta a) * \tau)(\cdot) = \sum_\rho (\delta a)(\rho, \tau)$$

$$= \sum_\rho (\rho a(\tau) - a(\rho\tau) + a(\rho))$$

$$= \sum_\rho \rho a(\tau)$$

$$= S_G(a(\tau))) = S_G((a * \tau)(\cdot))$$

$$= (\delta(a * \tau))(\cdot),$$

which proves the commutativity for $r = 1$.

Next let $r = 2$ and let $a(\sigma, \tau)$ be a 2-cochain. Then:

$$((\delta a) * \tau)(\sigma) = \sum_\rho (\delta a)(\sigma, \rho, \tau)$$

$$= \sum_\rho (\sigma a(\rho, \tau) - a(\sigma\rho, \tau) + a(\sigma, \rho\tau) - a(\sigma, \rho))$$

$$= \sum_\rho \sigma a(\rho, \tau) - a(\rho, \tau)$$

$$= (\sigma - e) \sum_\rho a(\rho, \tau)$$

$$= (\sigma - e)((a * \tau)(\cdot))$$

$$= (\delta(a * \tau))(\sigma),$$

which proves the commutativity relation for $r = 2$. For $r > 2$, the proof is entirely similar and is left to the reader.

From the commutativity relation, one obtains an induced homomorphism on the cohomology groups, namely

$$\varphi_\tau : \mathbf{H}^r(A) \to \mathbf{H}^{r-2}(A) \quad \text{for} \quad r \geqq 2.$$

For $r = 2$, we have to note that if the cochain $a(\sigma)$ is a coboundary, that is

$$a(\sigma) = (\sigma - e)b \quad \text{for some} \quad b \in A,$$

then $(a * \tau)(\cdot) = (\tau - e)b$ is in $I_G A$. Thus φ_τ is a morphism of functors. It is also a δ-morphism, i.e. φ_τ commutes with the coboundary associated with a short exact sequence. Since

$$\alpha \mapsto \alpha \cup \zeta_\tau$$

is also a δ-morphism of $\mathbf{H}^r$ to $\mathbf{H}^{r-2}$, to show that they are equal, it suffices to show that they coincide for $r = 1$, because of the uniqueness theorem.

Explicitly, we have to show that if $\alpha \in \mathbf{H}^1(A)$ is represented by the cocycle $a(\sigma)$, then $\alpha \cup \zeta_\tau$ is represented by $(a * \tau)(\cdot)$, that is

$$\alpha \cup \zeta_\tau = \varkappa(a(\tau)).$$

This is now clear, because of the diagram:

$$
\begin{array}{ccccc}
\varkappa(f) & \times & (\tau - e) & \longmapsto & (f(\tau - e)) \\
\Big\downarrow{\scriptstyle\delta} & & \Big\downarrow{\scriptstyle\delta^{-1}} & & \Big\downarrow{\scriptstyle\text{id}} \\
[\delta\varkappa(f) = -\alpha] & \times & \delta^{-1}(\tau - e) & \longmapsto & \alpha \cup \zeta_\tau.
\end{array}
$$

The coboundary on the left comes from Proposition 8.1. This concludes the proof.

Corollary 8.3. *If $\alpha \in \mathbf{H}^1(A)$ is represented by a standard co-cycle $a(\sigma)$, then for each $\tau \in G$ we have $a(\tau) \in A_{S_G}$ and*

$$\alpha \cup \zeta_\tau = \varkappa(a(\tau)) \in \mathbf{H}^{-1}(A).$$

If $\alpha \in \mathbf{H}^2(A)$ is represented by a standard cocycle $a(\sigma, \tau)$, then for each $\tau \in G$ we have $\sum_\rho a(\rho, \tau) \in A^G$, and

$$\alpha \cup \zeta_\tau = \varkappa\left(\sum_\rho a(\rho, \tau)\right) \in \mathbf{H}^0(A).$$

Corollary 8.4. *The duality between $\mathbf{H}^1(G, \mathbf{Q}/\mathbf{Z})$ and $\mathbf{H}^{-2}(G, \mathbf{Z})$ in the duality theorem is consistent with the identification of $\mathbf{H}^1(G, \mathbf{Q}/\mathbf{Z})$ with $\hat{G}$ and of $\mathbf{H}^{-2}(G, \mathbf{Z})$ with G/G^c.*

The above corollary pursues the considerations of Theorem 1.17, Chapter II, in the context of the cup product. We also obtain further commutativity relations in the next theorem.

Proposition 8.5. *Let U be a subgroup of G.*

(i) *For $\tau \in U, A \in \mathrm{Mod}(G), \alpha \in \mathbf{H}^r(G,A)$ we have*

$$\mathrm{tr}_G^U(\zeta_\tau \cup \mathrm{res}_U^G(\alpha)) = \zeta_\tau \cup \alpha.$$

(ii) *If U is normal, $m = (U:e)$, and $\alpha \in \mathbf{H}^r(G/U, A^U)$ with $r \geq 2$, then*

$$m \cdot \inf_G^{G/U}(\zeta_{\bar\tau} \cup \alpha) = \zeta_\tau \cup \inf_G^{G/U}(\alpha).$$

If $r = 2$, $m \cdot \inf_G^{G/U}$ is induced by the maps

$$B^G \to mB^G \quad and \quad m\mathbf{S}_U B \to \mathbf{S}_G B$$

for $B \in \mathrm{Mod}(G/U)$.

Proof. As to the first formula, since the transfer corresponds to the map induced by inclusion, we can apply directly the cup product formula from Theorem 3.1, that is

$$\mathrm{tr}(\alpha \cup \mathrm{res}(\beta)) = \mathrm{tr}(\alpha) \cup \beta.$$

One can also use the Nakayama maps, as follows. For τ fixed, we have two maps

$$\alpha \mapsto \zeta_\tau \cup \alpha \quad and \quad \alpha \mapsto \mathrm{tr}_G^U(\zeta_\tau \cup \mathrm{res}_U^G(\alpha))$$

which are immediately verified to be δ-morphisms of the cohomological functor $\mathbf{H}_G$ into $\mathbf{H}_G$ with a shift of 2 dimensions. To show that they are equal, it will suffice to do so in dimension 2. We apply Nakayama's formula. We use a coset decomposition $G = \bigcup \bar c U$ as usual. If f is a cocycle representing α, the first map corresponds to

$$f \mapsto \sum_{\rho \in G} f(\rho, \tau).$$

The second one is

$$\mathbf{S}_U^G \left(\sum_{\rho \in U} f(\rho, \tau) \right) = \sum_{c, \rho \in U} \bar c f(\rho, \tau).$$

Using the cocycle relation

$$f(\bar{c}, \rho) + f(\bar{c}\rho, \tau) - f(\bar{c}, \rho\tau) = \bar{c}f(\rho, \tau),$$

the desired equality falls out.

This method with the Nakayama map can also be used to prove the second part of the proposition, with the lifting morphism $\text{lif}_G^{G/U}$ replacing the inflation $\inf_G^{G/U}$. One sees that $m \cdot \text{lif}_G^{G/U}$ is a δ-morphism of $H_{G/U}$ to H_G on the category $\text{Mod}(G/U)$, and it will suffice to show that the δ-morphism.

$$\alpha \mapsto \zeta_\tau \cup \text{lif}_G^{G/U}(\alpha) \quad \text{and} \quad \alpha \mapsto m \cdot \text{lif}_G^{G/U}(\zeta_{\bar{\tau}} \cup \alpha)$$

coincide in dimension 2. This follows by using the Nakayama maps as in the first case.

If G is cyclic, then $\mathbf{H}^{-2}(G, \mathbf{Z})$ has a maximal generator, of order $(G : e)$, and -2 is a cohomological period. If σ is a generator of G, then for all $r \in \mathbf{Z}$, the map

$$\mathbf{H}^r(G, A) \to \mathbf{H}^{r-2}(G, A) \quad \text{given by} \quad \alpha \mapsto \zeta_\sigma \cup \alpha$$

is an isomorphism. Hence to compute the restriction, inflation, transfer, conjugation, we can use the commutativity formulas and the explicit formulas of Chapter II, §2.

Corollary 8.6. *Let G be cyclic and suppose $(U : e)$ divides the order of G/U. Then the inflation*

$$\inf_G^{G/U} : H^s(G/U, A^U) \to H^s(G, A)$$

is 0 for $s \geq 3$.

Proof. Write $s = 2r$ or $s = 2r + 1$ with $r \geq 1$. Let σ be a generator of G. By Proposition 8.5, we find

$$\zeta_\sigma \cup \inf(\alpha) = \inf(\zeta_{\bar{\sigma}} \cup \alpha).$$

But $\zeta_{\bar{\sigma}} \cup \alpha$ has dimension $s - 2$. By induction, its inflation is killed by $(U : e)^{r-1}$, from which the corollary follows.

The last theorem of this section summarizes some commutativities in the context of the cup product, extending the table from Chapter I, §4.

Theorem 8.7. *Let G be finite of order n. Let $A \in \mathrm{Mod}(G)$ and $\alpha \in \mathbf{H}^2(A) = \mathbf{H}^2(G, A)$. The following diagram is commutative.*

$$
\begin{array}{ccccc}
\mathbf{H}^0(A) & \times & \mathbf{H}^2(XZ) & \longrightarrow & \mathbf{H}^2(A) \\
\downarrow{\scriptstyle \cup\alpha} & & \downarrow{\scriptstyle \mathrm{id}} & & \downarrow{\scriptstyle \cup\alpha} \\
\mathbf{H}^{-2}(\mathbf{Z}) & \times & \mathbf{H}^2(\mathbf{Z}) & \longrightarrow & \mathbf{H}^0(\mathbf{Z}) = \mathbf{Z}/n\mathbf{Z} \\
\downarrow{\scriptstyle \mathrm{id}} & & \downarrow{\scriptstyle \delta} & & \downarrow{\scriptstyle \delta} \\
\mathbf{H}^{-2}(\mathbf{Z}) & \times & \mathbf{H}^1(\mathbf{Q}/\mathbf{Z}) & \longrightarrow & \mathbf{H}^{-1}(\mathbf{Z}/\mathbf{Z}) \\
\downarrow & & \downarrow & & \downarrow \\
G/G^c & \times & \hat{G} & \longrightarrow & (\mathbf{Z}/\mathbf{Z})_n
\end{array}
$$

The vertical maps are isomorphisms in the two lower levels. If A is a class module and α a fundamental element, i.e. a generator of $\mathbf{H}^2(A)$, then the cups with α on the first level are also isomorphisms.

Proof. The commutative on top comes from the fact that all elements have even dimension, and that one has commutativity of the cup product for even dimension. The lower commutativities are an old story. If A is a class module, we know that cupping with α gives an isomorphism, this being Tate's Theorem 7.1.

Remark. Theorem 8.7 gives, in an abstract context, the reciprocity isomorphism of class field theory. If G is abelian, then $G^c = e$ and $\mathbf{H}^0(A) = A^G/\mathbf{S}_G A$ is both isomorphic to G and dual to G. On one hand, it is isomorphic to G by cupping with α, and identifying $\mathbf{H}^{-2}(\mathbf{Z})$ with G. On the other hand, if χ is a character of G, i.e. a cocycle of dimension 1 in $\mathbf{Q}/\mathbf{Z}$, then the cupping

$$
\varkappa(a) \times \delta\chi \mapsto \varkappa(a) \cup \delta\chi
$$

gives the duality between $A^G/\mathbf{S}_G A$ and $\mathbf{H}^1(\mathbf{Q}/\mathbf{Z})$, the values being taken in $\mathbf{H}^2(A)$. The diagram expressed the fact that the identification of $\mathbf{H}^0(G, A)$ with G made in these two ways is consistent.

CHAPTER V
Augmented Products

§1. Definitions

In Tate's work a new cohomological operation was defined, satisfying properties similar to those of the cup product, but especially adjusted to the applications to class field theory and to the duality of cohomology on connection with abelian varieties. As usual here, we give the general setting which requires no knowledge beyond the basic elementary theory we are carrying out.

Let $\mathfrak{A}$ be an abelian bilinear category, and let H, E, F be three δ-functors on $\mathfrak{A}$ with values in the same abelian category $\mathfrak{B}$. For each integers r, s such that H^r, E^s are defined, we suppose that F^{r+s+1} is also defined. By a **Tate product**, we mean the data of two exact sequences

$$0 \to A' \xrightarrow{i} A \xrightarrow{j} A'' \to 0$$
$$0 \to B' \xrightarrow{i} B \xrightarrow{j} B'' \to 0$$

and two bilinear maps

$$A' \times B \to C \quad \text{and} \quad A \times B' \to C$$

coinciding on $A' \times B'$. Such data, denoted by (A, B, C), form a category in the obvious sense. An **augmented cupping**

$$H \times E \to F$$

110

associates to each Tate product a bilinear map

$$U_{aug} : H^r(A'') \times E^S(B'') \to F^{r+s+1}(C)$$

satisfying the following conditions.

ACup 1. The association is functorial, in other words, if
$u : (A, B, C) \to (\bar{A}, \bar{B}, \bar{C})$ is a morphism of a Tate product to
another, then the diagram

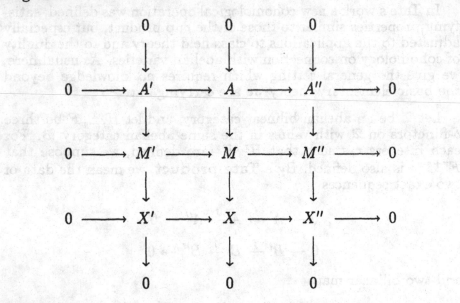

is commutative.

ACup 2. The augmented cupping satisfies the property of dimension shifting namely: Suppose given an exact and commutative diagram:

and two exact sequences

$$0 \to B' \to B \to B'' \to 0$$
$$0 \to C' \to M_C \to X_C \to 0,$$

as well as bilinear maps

$$A' \times B \to C \qquad\qquad B' \times B \to M_C$$
$$A \times B' \to C \qquad\qquad M \times B' \to M_C$$
$$X' \times B \to X_C$$
$$X \times B' \to X_C$$

which are compatible in the obvious sense, left to the reader, and coincide on $A' \times B'$ resp. $M' \times B'$, resp. $X' \times B'$, then

$$
\begin{array}{ccc}
H^r(X'') \times E^s(B'') & \longrightarrow & F^{r+s+1}(X_C) \\
\downarrow{\scriptstyle\delta} \qquad \downarrow{\scriptstyle\text{id}} & & \downarrow{\scriptstyle\delta} \\
H^{r+1}(A'') \times E^s(B'') & \longrightarrow & F^{r+s+2}(C)
\end{array}
$$

is commutative. Similarly, if we shift dimensions on E, then the similar diagram will have character -1.

If H is erasable by an erasing functor M which is exact, and whose cofunctor X is also exact, then we get a uniqueness theorem as in the previous situations.

Thus the agreed cupping behaves like the cup product, but in a little more complicated way. All the relations concerning restriction, transfer, etc. can be formulated for the augmented cupping, and are valid with similar proofs, based as before on the uniqueness theorem. For example, we have:

Proposition 1.1. *Let G be a group. Suppose given on H_G an augmented product. Let U be a subgroup of finite index. Given a Tate product (A, B, C), let*

$$\alpha'' \in H^r(G, A'')$$

and

$$\beta'' \in H^s(U, B'').$$

Then

$$\mathrm{tr}_G^U(\mathrm{res}_U^G(\alpha'') \cup_{\mathrm{aug}} \beta'') = \alpha'' \cup_{\mathrm{aug}} \mathrm{tr}_G^U(\beta'').$$

Proof. Both sides of the above equality define an augmented cupping $H_G \times H_U \to H_G$, these cohomological functors being taken on the multilinear category $\mathrm{Mod}(G)$. They coincide in dimension $(0, 0)$ and 1, as one determines by an explicit computation, so the general uniqueness theorem applies.

Proposition 1.2. *For $\alpha'' \in H^r(U, A)$ and $\beta'' \in H^s(U, B'')$ and $\sigma \in G$, we have*

$$\sigma_*(\alpha'' \cup_{\mathrm{aug}} \beta'') = \sigma_*\alpha'' \cup_{\mathrm{aug}} \sigma_*\beta''.$$

Similarly, if U is normal in G and $\alpha'' \in H^r(G/U, A''^U)$, $\beta'' \in H^s(G/U, B''^U)$, we have for the inflation

$$\inf(\alpha'' \cup_{\mathrm{aug}} \beta'') = \inf(\alpha'') \cup_{\mathrm{aug}} \inf(\alpha'').$$

Of course, the above statements hold for the special functor $\mathbf{H}$. when G is finite, except when we deal with inflation.

We make the augmented product more explicit in dimensions $(-1, 0)$ and 0, as well as $(0, 0)$ and 1, for the special functor $\mathbf{H}_G$.

Dimensions $(-1, 0)$ and 0. We are given two exact sequences

$$0 \to A' \xrightarrow{i} A \xrightarrow{j} A'' \to 0$$

$$0 \to B' \xrightarrow{i} B \xrightarrow{j} B'' \to 0$$

as well as a Tate product, that is bilinear maps in $\mathrm{Mod}(G)$:

$$A' \times B \to C \quad \text{and} \quad A \times B' \to C$$

coinciding on $A' \times B'$. We then define the augmented product by

$$\varkappa(a'') \cup_{\mathrm{aug}} \varkappa(b'') = \varkappa(a'b - ab'),$$

where a', b' are determined as follows. We choose $a \in A$ such that $ja = a''$ and $b \in B$ such that $jb = b''$. Then a', b' are uniquely determined by the conditions

$$ia' = \mathbf{S}_G(a) \quad \text{and} \quad ib' = \mathbf{S}_G(b).$$

Dimensions $(0, 0)$ and 1. We define

$$\varkappa(a'') \cup_{\mathrm{aug}} \varkappa(b'') = \text{cohomology class of the cocycle} \quad a'_\sigma b + ab'_\sigma,$$

where the cocycle a'_σ is determined by the formulas

$$ja = a'' \quad \text{and} \quad ia''_\sigma = \sigma a - a,$$

and similarly for b'.

§2. Existence

The existence is given in a way similar to that of the cup product. We shall be very brief. First an abstract statement:

Theorem 2.1. *Let $\mathfrak{A}$ be a multilinear abelian category, and suppose given an exact bilinear functor $A \mapsto Y(A)$ from $\mathfrak{A}$ into the bilinear category of complexes in an abelian category $\mathfrak{B}$. Then the corresponding cohomological functor H on $\mathfrak{A}$ has a structure of augmented cup functor, in the manner described below.*

Recall from Chapter IV, §2 that we already described how the category of complexes forms a bilinear category. For the application to the augmented cup functor, suppose that (A, B, C) is a Tate product. We want to define a bilinear map

$$H^r(A'') \times H^s(B'') \to H^{r+s+1}(C).$$

We do so by a bilinear map defined on the cochains as follows. Let α'' and β'' be cohomology classes in $H^r(A'')$ and $H^s(B'')$ respectively, and let f'', g'' be representative cochains in $Y^r(A), Y^s(B)$ respectively, so that $jf = f''$ and $jg = g''$. We view i as an inclusion, and we let

$$h = \delta f . g + (-1)^r f . \delta g,$$

the products on the right being the Tate product. Then we define $\alpha'' \cup_{\text{aug}} \beta''$ to be the cohomology class of h.

One verifies tediously that this class is independent of the choices made in its construction, and one also proves the dimension shifting property, which is actually a pain, which we do not carry out.

§3. Some properties

Theorem 3.1. *Let the notation be as in Theorem 2.1 with a Tate product (A, B, C). Then the squares in the following diagram from left to right are commutative, resp. of character $(-1)^r$, resp. commutative.*

$$
\begin{array}{ccccccc}
H^r(A') & \to & H^r(A) & \to & H^r(A'') & \xrightarrow{\delta} & H^{r+1}(A') \\
\times & & \times & & \times & & \times \\
H^s(B) & \leftarrow & H^{s+1}(B') & \xleftarrow{\delta} & H^s(B'') & \leftarrow & H^s(B) \\
\cup\downarrow & & \cup\downarrow & & \downarrow\cup_{\text{aug}} & & \downarrow\cup \\
H^{r+s+1}(C) & \to & H^{r+s+1}(C) & \to & H^{r+s+1}(C) & \to & H^{r+s+1}(C)
\end{array}
$$

114

The morphisms on the bottom line are all the identity.

Proof. The result follows immediately from the definition of the cup and augmented cup in terms of cochain representatives, both for the ordinary cup and the augmented cup.

The next property arose in Tate's application of cohomology theory to abelian varieties. See Chapter X.

Theorem 3.2. *Let the multilinear categories be those of abelian groups. Let m be an integer ≥ 1, and suppose that the following sequences are exact:*

$$0 \to A''_m \to A'' \xrightarrow{m} A'' \to 0$$
$$0 \to B''_m \to B'' \xrightarrow{m} B'' \to 0.$$

Given a Tate product (A, B, C), one can define a bilinear map

$$A''_m \times B''_m \to C$$

as follows. Let $a'' \in A''_m$ and $b'' \in B''_m$. Choose $a \in A$ and $b \in B$ such that $ja = a''$ and $jb = b''$. We define

$$\langle a'', b'' \rangle = ma.b - a.mb.$$

Then the map $(a'', b'') \mapsto \langle a'', b'' \rangle$ is bilinear.

Proof. Immediate from the definitions and the hypothesis on a Tate product.

Theorem 3.3. *Let (A, B, C) be a Tate product in a multilinear abelian category of abelian groups. Notation as in Theorem 3.2, we have a diagram of character $(-1)^{r-1}$:*

$$
\begin{array}{ccc}
H^r(A''_m) & \longrightarrow & H^r(A'') \\
\times & & \times \\
H^{s+1}(B''_m) & \xleftarrow{\ \delta\ } & H^s(B'') \\
\cup \downarrow & & \downarrow \cup_{\text{aug}} \\
H^{r+s+1}(C) & \xrightarrow[\text{id}]{} & H^{r+s+1}(C)
\end{array}
$$

Note that the coboundary map in the middle is the one associated with the exact sequence involving B''_m and B'' in Theorem 3.2. The cup product on the left is the one obtained from the bilinear map as in Theorem 3.2.

CHAPTER VI
Spectral Sequences

We recall some definitions, but we assume that the reader knows the material of *Algebra*, Chapter XX, §9 on spectral sequences, their basic constructions and more elementary properties.

§1. Definitions

Let $\mathfrak{A}$ be an abelian category and A an object in $\mathfrak{A}$. A **filtration** of A consists in a sequence

$$F = F^0 \supset F^1 \supset F^2 \supset \ldots \supset F^n \supset F^{n+1} = 0.$$

If F is given with a differential (i.e. endomorphism) d such that $d^2 = 0$, we also assume that $dF^p \subset F^p$ for all $p = 0, \ldots, n$, and one then calls F a **filtered differential object.** We define the graded object

$$\mathrm{Gr}(F) = \bigoplus_{p \geqq 0} \mathrm{Gr}^p(F) \quad \text{where} \quad \mathrm{Gr}^p(F) = F^p/F^{p+1}.$$

We may view $\mathrm{Gr}(F)$ as a complex, with a differential of degree 0 induced by d itself, and we have the homology $H(\mathrm{Gr}^p F)$.

Filtered objects form an additive category, which is not necessarily abelian. The family $\mathrm{Gr}(A)$ defines a covariant functor on the category of filtered objects.

A **spectral sequence** in A is a family $E = (E_r^{p,q}, E^n)$ consisting of:

(1) Objects $E_r^{p,q}$ defined for integers p, q, r with $r \geq 2$.

(2) Morphisms $d_r^{p,q} : E_r^{p,q} \to E_r^{p+r,q-r+1}$ such that

$$d_r^{p+r,q-r+1} \circ d_r^{p,q} = 0.$$

(3) Isomorphisms

$$\alpha_r^{p,q} : \mathrm{Ker}(d_r^{p,q})/\mathrm{Im}(d_r^{p-r,q+r-1}) \to E_{r+1}^{p,q}.$$

(4) Filtered objects E^n in A defined for each integer n.

We suppose that for each pair (p, q) we have $d_r^{p,q} = 0$ and $d_r^{p-r,q+r-1} = 0$ for r sufficiently large. It follows that $E_r^{p,q}$ is independent of r for r sufficiently large, and one denotes this object by $E_\infty^{p,q}$. We assume in addition that for n fixed, $F^p(E^n) = E^n$ for p sufficiently small, and is equal to 0 for p sufficiently large.

Finally, we suppose given:

(5) Isomorphisms $\beta^{p,q} : E_\infty^{p,q} \to \mathrm{Gr}^p(E^{p+q})$.

The family $\{E^n\}$, with filtration, is called the **abutment** of the spectral sequence E, and we also say that E **abuts** to $\{E^n\}$ or converges to $\{E^n\}$.

By general principles concerning structures defined by arrows, we know that spectral sequences in $\mathfrak{A}$ form a category. Thus a morphism $u : E \to E'$ of a spectral sequence into another consists in a system of morphisms.

$$u_r^{p,q} : E_r^{p,q} \to E_r^{p,q} \quad \text{and} \quad u^n : E^n \to E''^n.$$

compatible with the filtrations, and commuting with the morphisms $d_r^{p,q}, \alpha_r^{p,q}$ and $\beta_r^{p,q}$. Spectral sequences in $\mathfrak{A}$ then form an additive category, but not an abelian category.

A **spectral functor** is an additive functor on an abelian category, with values in a category of spectral sequences.

We refer to *Algebra*, Chapter XX, §9 for constructions of spectral sequences by means of double complexes.

A spectral sequence is called **positive** if $E_r^{p,q} = 0$ for $p < 0$ and $q < 0$. This being the case, we get:

$$E_r^{p,q} \approx E_\infty^{p,q} \qquad \text{for } r > \sup(p, q+1)$$
$$E^n = 0 \qquad \text{for } n < 0$$
$$F^m(E^n) = 0 \qquad \text{if } m > n$$
$$F^m(E^n) = E^n \qquad \text{if } m \leqq 0.$$

In what follows, we assume that all spectral sequences are positive.

We have inclusions

$$E^n = F^0(E^n) \supset F^1(E^n) \supset \ldots \supset F^n(E^n) \supset F^{n+1}(E^n) = 0.$$

The isomorphisms

$$\beta^{0,n} : E_\infty^{0,n} \to \mathrm{Gr}^0(E^n) = F^0(E^n)/F^1(E^n) = E^n/F^1(E^n)$$
$$\beta^{n,0} : E_\infty^{n,0} \to \mathrm{Gr}^n(E^n) = F^n(E^n)$$

will be called the **edge**, or **extreme**, isomorphisms of the spectral sequence.

Theorem 9.6 of *Algebra*, Chapter XX, shows how to obtain a spectral sequence from a composite of functors under certain conditions, the **Grothendieck spectral sequence**. We do not repeat this result here, but we shall use it in the next section.

§2. The Hochschild-Serre spectral sequence

We now apply spectral sequence to the cohomology of groups. Let G be a group and H_G the cohomological functor on $\mathrm{Mod}(G)$. Let N be a normal subgroup of G. Then we have two functors:

$$A \mapsto A^N \text{ of } \mathrm{Mod}(G) \text{ into } \mathrm{Mod}(G/N)$$

$$B \mapsto B^{G/N} \text{ of } \mathrm{Mod}(G/N) \text{ into Grab (abelian groups).}$$

Composing these functors yields $A \mapsto A^G$. Therefore, we obtain the Grothendieck spectral sequence associated to a composite of functors, such that for $A \in \mathrm{Mod}(G)$:

$$E_2^{p,q}(A) = H^p(G/N, H^q(N, A)),$$

with G/N acting on $H^q(N, A)$ by conjugation as we have seen in Chapter II, §1. Furthermore, this spectral functor converges to

$$E^n(A) = H^n(G, A).$$

One now has to make explicit the edge homomorphisms. First we have an isomorphism

$$\beta^{0,n} : E_\infty^{0,n}(A) \to H^n(G, A)/F^1(H^n(G, A)),$$

where F^1 denotes the first term of the filtration. Furthermore

$$E_2^{0,n}(A) = H^n(N, A)^{G/N}$$

and $E_\infty^{0,n}(A)$ is a subgroup of $E_2^{0,n}(A)$, taking into account that the spectral sequence is positive. Hence the inverse of $\beta^{0,n}$ yields a monomorphism of $H^n(G, A)/F^1(H^n(G, A))$ into $H^n(N, A)$, and induces a homomorphism

$$H^n(G, A) \to H^n(N, A).$$

Proposition 2.1. *This homomorphism induced by the inverse of $\beta^{0,n}$ is the restriction.*

Proof. This is a routine tedious verification of the edge homomorphism in dimension 0, left to the reader.

In addition, we have an isomorphism

$$\beta^{n,0} : E_\infty^{n,0}(A) \to F^n(H^n(G, A)),$$

whose image is a subgroup of $H^n(G, A)$. Dually to what we had previously, $E_\infty^{n,0}(A)$ is a factor group of $E_2^{n,0}(A) = H^n(G/N, A^N)$. Composing the canonical homomorphism coming from the $d_r^{n,0}$ and $\beta^{n,0}$, we find a homomorphism

$$H^n(G/N, A^N) \to H^n(G, A).$$

Proposition 2.2. *This homomorphism is the inflation.*

Proof. Again omitted.

Besides the above edge homomorphisms, we can also make the spectral sequence more explicit, both in the lowest dimension and under other circumstances, as follows.

Theorem 2.3. *Let G be a group and N a normal subgroup. Then for $A \in Mod(G)$ we have an exact sequence:*

$$0 \to H^1(G/N, A^N) \xrightarrow{\text{inf}} H^1(G, A) \xrightarrow{\text{res}} H^1(N, A)^{G/N} \xrightarrow{d_2}$$
$$\xrightarrow{d_2} H^2(G/N, A^N) \xrightarrow{\text{inf}} H^2(G, A).$$

The homomorphism d_2 in the above sequence is called the **transgression**, and is denoted by tg, so

$$\text{tg} : H^1(N, A)^{G/N} \to H^2(G/N, A^N).$$

This map tg can be defined in higher dimensions under the following hypothesis.

Theorem 2.4. *If $H^r(N, A) = 0$ for $1 \leq r < s$, then we have an exact sequence:*

$$0 \to H^s(G/N, A^N) \xrightarrow{\text{inf}} H^s(G, A) \xrightarrow{\text{res}} H^s(N, A)^{G/N} \xrightarrow{\text{tg}}$$
$$\xrightarrow{\text{tg}} H^{s+1}(G/N, A^N) \xrightarrow{\text{inf}} H^{s+1}(G, A).$$

For computations, it is useful to describe tg in dimension 1 in terms of cochains, so we consider tg as in Theorem 2.3, in dimension 1, and we have:

An element $\alpha \in H^2(G/N, A^N)$ can be written as

$$\alpha = \text{tg}(\beta) \text{ with } \beta \in H^1(N, A)^{G/N}$$

if and only if there exists a cochain $f \in C^1(G, A)$ such that:

1. The restriction of f to N is a 1-cocycle representing β.

2. We have $\delta f = $ inflation of a 2-cocycle representing α.

In case many groups $H^r(N, A)$ are trivial, the spectral sequence gives isomorphisms and exact sequences as in the next two theorems.

Theorem 2.5. *Suppose $H^r(N, A) = 0$ for $r > 0$. Then*

$$\alpha_2^{p,0} : H^p(G/N, A^N) \longrightarrow H^p(G, A)$$

is an isomorphism for all $p \geq 0$.

The hypothesis in Theorem 2.5 means that all points of the spectral sequence are 0 except those of the bottom line. Furthermore:

Theorem 2.6. *Suppose that $H^r(N, A) = 0$ for $r > 1$. Then we have an infinite exact sequence:*

$$0 \to H^1(G/N, H^0(N,A)) \to H^1(G,A) \to H^0(G/N, H^1(N,A)) \to$$

$$\xrightarrow{d_2} H^2(G/N, H^0(N,A)) \to H^2(G,A) \to H^1(G/N, H^1(N,A)) \to$$

$$\xrightarrow{d_2} H^3(G/N, H^0(N,A)) \to H^3(G,A) \to H^2(G/N, H^1(N,A)) \to$$

The hypothesis in Theorem 2.6 means that all the points of the spectral sequence are 0 except those of the two bottom lines.

§3. Spectral sequences and cup products

In this section we state two theorems where cup products occur within spectral sequences. We deal with the multilinear category Mod(G), a normal subgroup N of G, and the Hochschild-Serre spectral sequence.

Theorem 3.1. *The spectral sequence is a cup functor (in two dimensions) in the following sense. To each bilinear map*

$$A \times B \to C$$

there is a cupping determined functorially

$$E_r^{p,q}(A) \times E_r^{p',q'}(B) \longrightarrow E_r^{p+p',q+q'}(C)$$

such that for $\alpha \in E_r^{p,q}(A)$ and $\beta \in E_r^{p',q'}(B)$ we have

$$d_r(\alpha \cdot \beta) = (d_r\alpha) \cdot \beta + (-1)^{p+q}\alpha \cdot (d_r\beta).$$

If we denote by $\cup$ the usual cup product, then for $r = 2$

$$\alpha \cup \beta = (-1)^{q'p}\alpha \cdot \beta.$$

The cupping is induced by the bilinear map

$$H^q(N,A) \times H^{q'}(N,B) \to H^{q+q'}(N,C).$$

Finally, suppose G is a finite group, and $B \in \mathrm{Mod}(G)$. We have an exact sequence with arrows pointing to the left:

$$0 \leftarrow B^G/S_{G/N}B^N \xleftarrow{\mathrm{can}} B^G \xleftarrow{\mathrm{S}_{G/N}} B^N/I_{G/N}B^N \xleftarrow{\mathrm{inc}} B^N_{S_{G/N}}/I_{G/N}B^N \leftarrow 0,$$

or in other words

$$0 \leftarrow \mathbf{H}^0(G/N,B^N) \leftarrow H^0(G,B) \leftarrow H_0(G/N,B^N) \leftarrow \mathbf{H}^{-1}(G/N,B^N) \leftarrow 0$$

This exact sequence is dual to the inflation-restriction sequence, in the following sense.

Theorem 3.2. *Let G be a group, U a normal subgroup of finite index, and (A,B,C) a Tate product in $\mathrm{Mod}(G)$. Suppose that (A^U, B^U, C^U) is also a Tate product. Then the following diagram is commutative:*

$$
\begin{array}{ccccc}
H^1(G/U, A''^U) & \xrightarrow{\mathrm{inf}} & H^1(G, A'') & \xrightarrow{\mathrm{res}} & H^1(U, A'') \\
\times & & \times & & \times \\
H^0(G/U, B''^U) & \xleftarrow{\mathrm{can}} & H^0(G, B'') & \xleftarrow{\mathrm{S}^U_G} & H^0(U, B'') \\
\downarrow{\scriptstyle \cup_{\mathrm{aug}}} & & \downarrow & & \downarrow \\
H^2(G/U, C^U) & \xrightarrow{\mathrm{inf}} & H^2(G, C) & \xrightarrow{\mathrm{tr}} & H^2(U, C).
\end{array}
$$

with $0 \leftarrow$ at the left of the middle row.

The two horizontal sequences on top are exact.

CHAPTER VII
Groups of Galois Type
(Unpublished article of Tate)

§1. Definitions and elementary properties

We consider here a new category of groups and a cohomological functor, obtained as limits from finite groups.

A topological group G will be said to be of **Galois type** if it is compact, and if the normal open subgroups form a fundamental system of neighborhoods of the identity e. Since such a group is compact, it follows that every open subgroup is of finite index in G, and is therefore closed.

Let S be a closed subgroup of G (no other subgroups will ever be considered). *Then S is the intersection of the open subgroups U containing S.* Indeed, if $\sigma \in G$ and $\sigma \notin S$, we can find an open normal subgroup U of G such that $U\sigma$ does not intersect S, and so $US = SU$ does not contain σ. But US is open and contains S, whence the assertion.

We observe that every closed subgroup of finite index is also open. Warning: There may exist subgroups of finite index which are not open or closed, for instance if we take for G the invertible power series over a finite field with p elements, with the usual topology of formal power series. The factor group G/G^p is a vector space over

$\mathbf{F}_p$, one can choose an intermediate subgroup of index p which is not open.

Examples of groups of Galois type come from Galois groups of infinite extensions in field theory, p-adic integers, etc.

Groups of Galois type form a category, the morphisms being the continuous homomorphisms. This category is stable under the following operations:

1. Taking factor groups by closed normal subgroups.
2. Products.
3. Taking closed subgroups.
4. Inverse limits (which follows from conditions 2 and 3).

Finite groups are of Galois type, and consequently every inverse limit of finite groups is of Galois type. Conversely, every group of Galois type is the inverse limit of its factor groups G/U taken over all open normal subgroups. Thus one often says that a Group of Galois type is **profinite**.

The following result will allow us to choose coset representatives as in the theory of discrete groups, which is needed to make the cohomology of finite groups go over formally to the groups of Galois type.

Proposition 1.1. *Let G be of Galois type, and let S be a closed subgroup of G. Then there exists a continuous section*

$$G/S \to G,$$

i.e. one can choose representatives of left cosets of S in G in a continuous way.

Proof. Consider pairs (T, f) formed by a closed subgroup T and a continuous map $f : G/S \to G/T$ such that for all $x \in G$ the coset $f(xS) = yT$ is contained in xS. We define a partial order by putting $(T, F) \leq (T_1, f_1)$ if $T \subset T_1$ and $f_1(xS) \subset f(xS)$. We claim that these pairs are then inductively ordered. Indeed, let $\{(T_i, f_i)\}$ be a totally ordered subset. Let $T = \bigcap_i T_i$. Then T is a closed subgroup of G. For each $x \in G$, the intersection

$$\bigcap f_i(xS)$$

is a coset of T_i, and is closed in S. Indeed, the finite intersection of such cosets $f_i(xS)$ is not empty because of the hypothesis on the maps f_i. The intersection $\bigcap f_i(xS)$ taken over all indices i is therefore not empty. Let y be an element of this intersection. Then by definition, $yT_i = f_i(xS)$ for all i, and hence $yT \subset f_i(xS)$ for all i. We define $f(xS) = yT$. Then $f(xS) \subset xS$.

The projective limit of the homogeneous spaces G/T_i is then canonically isomorphic to G/T, as one verifies immediately by the compactness of the objects involved. Hence the continuous sections $G/S \to G/T_i$ which are compatible can be lifted to a continuous section $G/S \to G/T$. By Zorn's lemma, we may suppose G/T is maximal, in other words, T is minimal. We have to show that $T = e$.

In other words, with the subgroup S given as at the beginning, if $S \neq e$ it will suffice to find $T \neq S$ and T open in S, closed in G, such that we can find a section $G/S \to S/T$. Let U be a normal open in G, $U \cap S \neq S$, and put $U \cap S = T$. If $G = \bigcup x_i US$ is a coset decomposition, then the map

$$x_i uS \mapsto x_i uT \quad \text{for} \quad u \in U$$

gives the desired section. This concludes the proof of Proposition 1.1.

We shall now extend to closed subgroups of groups of Galois type the notion of index. By a **supernatural number**, we mean a formal product

$$\prod p^{n_p}$$

taken over all primes p, the exponents n_p being integers ≥ 0 or ∞. One multiplies such products by adding the exponents, and they are ordered by divisibility in the obvious manner. The sup and inf of an arbitrary family of such products exist in the obvious way. If S is a closed subgroup of G, then we define the **index** $(G : S)$ to be equal to the supernatural number

$$(G : S) = \operatorname*{l.c.m.}_{V} (G : V),$$

the least common multiple l.c.m. being taken over open subgroups V containing S. Then one sees that $(G : S)$ is a natural number if and only if S is open. One also has:

Proposition 1.2. *Let $T \subset S \subset G$ be closed subgroups of G. Then*
$$(G : S)(S : T) = (G : T).$$

If (S_i) is a decreasing family of closed subgroups of G, then
$$(G : \bigcap_i S_i) = \text{l.c.m.}_i \ (G : S_i).$$

Proof. Let us prove the first assertion. Let m, n be integers ≥ 1 such that m divides $(G : S)$ and n divides $(S : T)$. We can find two open subgroups U, V of G such that $U \supset S, V \supset T$, m divides $(G : U)$ and n divides $(S : V \cap S)$. We have

$$(G : S \cap V) = (G : U)(U : U \cap V).$$

But there is an injection $S/(V \cap S) \to U/(U \cap V)$ of homogeneous spaces. By definition, one sees that mn divides $(G : T)$, and it follows that
$$(G : S)(S : T) \quad \text{divides} \quad (G : T).$$

One shows the converse divisibility by observing that if $U \supset T$ is open, then

$$(G : U) = (G : US)(US : U) \quad \text{and} \quad (US : U) = (S : S \cap U),$$

whence $(G : T)$ divides the product. This proves the first assertion of Proposition 1.2. The second assertion is proved by applying the first.

Let p be a fixed prime number. We say that G is a **p-group** if $(G : e)$ is a power of p, which is equivalent to saying that G is the inverse limit of finite p-groups. We say that S is a **Sylow p-subgroup** of G if S is a p-group and $(G : S)$ is prime to p.

Proposition 1.3. *Let G be a group of Galois type and p a prime number. Then G has a p-Sylow subgroup, and any two such subgroups are conjugate. Every closed p-subgroup S of G is contained in a p-Sylow subgroup.*

Proof. Consider the family of closed subgroups T of G containing S and such that $(G : T)$ is prime to p. It is partially ordered by descending inclusion, and it is actually inductively ordered since the intersection of a totally ordered family of such subgroups contains S and has index prime to p by Proposition 1.2. Hence the family contains a minimal element, say T. Then T is a p-group. Otherwise, there would exist an open normal subgroup U of G such that $(T : T \cap U)$ is not a p-power. Taking a Sylow subgroup of the finite group $T/(T \cap U) = TU/U$, for a prime number $\neq p$, once can find an open subgroup V of GH such that $(T : T \cap V)$ is prime to p, and hence $(S : S \cap V)$ is also prime to p. Since S is a p-group, one must have $S = S \cap V$, in other words $V \supset S$, and hence also $T \cap V \supset S$. This contradicts the minimality of T, and shows that T is a p-group of index prime to p, in other words, a p-Sylow subgroup.

Next let S_1, S_2 be two p-Sylow subgroups of G. Let $S_1(U)$ be the image of S under the canonical homomorphism $G \to G/U$ for U open normal in G. Then

$$(G/U : S_1 U/U) \quad \text{divides} \quad (G : S_1 U),$$

and is therefore prime to p. Hence $S_1(U)$ is a Sylow subgroup of G/U. Hence there exists an element $\sigma \in G$ such that $S_2(U)$ is conjugate to $S_1(U)$ by $\sigma(U)$. Let F_U be the set of such σ. It is a closed subset, and the intersection of a finite number of F_U is not empty, again because the conjugacy theorem is known for finite groups. Let σ be in the intersection of all F_U. Then S_1^σ and S_2 have the same image by all homomorphisms $G \to G/U$ for U open normal in G, whence they are equal, thus proving the theorem.

Next, we consider a new category of modules, to take into account the topology on a group of Galois type G. Let $A \in \mathrm{Mod}(G)$ be an ordinary G-module. Let

$$A_0 = \bigcup A^U,$$

the union being taken over all normal open subgroups U. Then A_0 is a G-submodule of A and $(A_0)_0 = A_0$. We denote by $\mathrm{Galm}(G)$ the category of G-modules A such that $A = A_0$, and call it the category of **Galois modules**. Note that if we give A the discrete topology, then $\mathrm{Galm}(G)$ is the subcategory of G-modules such that G operates continuously, the orbit of each element being finite, and the isotropy group being open. The morphisms in $\mathrm{Galm}(G)$ are the ordinary G-homomorphisms, and we still write $\mathrm{Hom}_G(A, B)$ for $A, B \in \mathrm{Galm}(G)$. Note that $\mathrm{Galm}(G)$ is an abelian category (the kernel and cokernel of a homomorphism of Galois modules are again Galois modules).

Let $A \in \mathrm{Galm}(G)$ and $B \in \mathrm{Mod}(G)$. Then

$$\mathrm{Hom}_G(A, B) = \mathrm{Hom}_G(A, B_0)$$

because the image of A by a G-homomorphism is automatically contained in B_0. From this we get the existence of enough injectives in $\mathrm{Galm}(G)$, as follows:

Proposition 1.4. *Let G be of Galois type. If $B \in \mathrm{Mod}(G)$ is injective in $\mathrm{Mod}(G)$, then B_0 is injective in $\mathrm{Galm}(G)$. If $A \in \mathrm{Galm}(G)$, then there exists an injective $M \in \mathrm{Galm}(G)$ and a monomorphism $u : A \to M$.*

Thus we can define the derived functor of $A \mapsto A^G$ in $\mathrm{Galm}(G)$, and we denote this functor again by H_G, so

$$H^0(G, A) = H_G^0(A) = A^G$$

as before.

Proposition 1.5. *Let G be of Galois type and N a closed normal subgroup of G. Let $A \in \mathrm{Galm}(G)$. If A is injective in $\mathrm{Galm}(G)$, then A^N is injective in $\mathrm{Galm}(G/N)$.*

Proof. If $B \in \mathrm{Galm}(G/N)$, we may consider B as an object of $\mathrm{Galm}(G)$, and we obviously have

$$\mathrm{Hom}_G(B, A) = \mathrm{Hom}_{G/N}(B, A^N)$$

because the image of B by a G-homomorphism is automatically contained in A^N. Considering these Hom as functors of objects B in $\mathrm{Galm}(G/N)$, we see at once that the functor on the right of the equality is exact if and only if the functor on the left is exact.

§2. Cohomology

(a) Existence and uniqueness. One can define the cohomology by means of the standard complex. For $A \in \mathrm{Galm}(G)$, let us put:

$$C^r(G, A) = 0 \quad \text{if} \quad r = 0$$
$$C^0(G, A) = A$$
$$C^r(G, A) = \text{groups of maps} \quad f : G^r \to A \quad (\text{for } r > 0),$$

continuous for the discrete topology on A.

We define the coboundary

$$\delta_r : C^r(G, A) \to C^{r+1}(G, A)$$

by the usual formula as in Chapter I, and one sees that $C(G, A)$ is a complex. Furthermore:

Proposition 2.1. *The functor $A \mapsto C(G, A)$ is an exact functor of $\mathrm{Galm}(G)$ into the category of complexes of abelian groups.*

Proof. Let

$$0 \to A' \to A \to A'' \to 0$$

be an exact sequence in $\mathrm{Galm}(G)$. Then the corresponding sequence of standard complexes is also exact, the surjectivity on the right being due to the fact that modules have the discrete topology, and that every continuous map $f : G^r \to A''$ can therefore be lifted to a continuous map of G^r into A.

By Proposition 1.5, we therefore obtain a δ-functor defined in all degrees $r \in \mathbf{Z}$ and 0 for $r < 0$, such that in dimension 0 this functor is $A \mapsto A^G$. We are going to see that this functor vanishes on injectives for $r > 0$, and hence by the uniqueness theory, that this δ-functor is isomorphic to the derived functor of $A \mapsto A^G$, which we denoted by H_G.

Theorem 2.2. *Let G be a group of Galois type. Then the cohomological functor H_G on $\mathrm{Galm}(G)$ is such that:*

$$H^r(G, A) = 0 \quad \text{for} \quad r > 0.$$
$$H^0(G, A) = A^G.$$
$$H^r(G, A) = 0 \quad \text{if A is injective,} \quad r > 0.$$

Proof. Let $f(\sigma_1, \ldots, \sigma_r)$ be a standard cocycle with $r \geq 1$. There exists a normal open subgroup U such that f depends only on cosets of U. Let A be injective in $\mathrm{Galm}(G)$. There exists an open normal subgroup V of G such that all the values of f are in A^V because f takes on only a finite number of values. Let $W = U \cap V$. Then f is the inflation of a cocycle $\bar{f}$ of G/W in A^W. By Proposition 1.5, we know that A^W is injective in $\mathrm{Mod}(G/W)$. Hence $\bar{f} = \delta \bar{g}$ with a cochain $\bar{g}$ of G/W in A^W, and so $f = \delta g$ if g is the inflation of $\bar{G}$ to G. Moreover, g is a continuous cochain, and so we have shown that f is a coboundary, hence that $H^r(G, A) = 0$.

In addition, the above argument also shows:

Theorem 2.3. *Let G be a group of Galois type and $A \in \mathrm{Galm}(G)$. Then*

$$H^r(G, A) \approx \mathrm{dir} \lim H^r(G/U, A^U),$$

the direct limit dir lim *being taken over all open normal subgroups U of G, with respect to inflation. Furthermore, $H^r(G, A)$ is a torsion group for $r > 0$.*

Thus we see that we can consider our cohomological functor H_G in three ways: the derived functor, the limit of cohomology groups of finite groups, and the homology of the standard complex.

For the general terminology of direct and inverse limits, cf. *Algebra*, Chapter III, §10, and also Exercises 16 - 26. We return to such limits in (c) below.

Remark. Let G be a group of Galois type, and let $\in \mathrm{Galm}(G)$. If G acts trivially on A, then similar to a previous remark, we have

$$H^1(G, A) = \mathrm{cont} \, \mathrm{hom}(G, A),$$

i.e. $H^1(G, A)$ consists of the continuous homomorphism of G into A. One sees this immediately from the standard cocycles, which are characterized by the condition

$$f(\sigma) + f(\tau) = f(\sigma\tau)$$

in the case of trivial action. In particular, take $A = \mathbf{F}_p$. Then as in the discrete case, we have:

Let G be a p-group of Galois type. If $H^1(G, \mathbf{F}_p) = 0$ then $G = e$, i.e. G is trivial.

Indeed, if $G \neq e$, then one can find an open subgroup U such that G/U is a finite p-group $\neq e$, and then one can find a non-trivial homomorphism $\lambda : G/U \to \mathbf{F}_p$ which, composed with the canonical homomorphism $G \to G/U$ would give rise to a non-trivial element of $H^1(G, \mathbf{F}_p)$.

(b) **Changing the group**. The theory concerning changes of groups is done as in the discrete case. Let $\lambda : G' \to G$ be a continuous homomorphism of a group of Galois type into another. Then λ gives rise to an exact functor

$$\Phi_\lambda : \mathrm{Galm}(G) \to \mathrm{Galm}(G'),$$

meaning that every object $A \in \mathrm{Galm}(G)$ may be viewed as a Galois module of G'. If

$$\varphi : A \to A'$$

is a morphism in $\mathrm{Galm}(G')$, with $A \in \mathrm{Galm}(G), A' \in \mathrm{Galm}(G')$, with the abuse of notation writing A instead of $\Phi_\lambda(A)$, the pair (λ, φ) determines a homomorphism

$$H^r(\lambda, \varphi) = (\lambda, \varphi)_* : H^r(G, A) \to H^r(G', A'),$$

functorially, exactly as for discrete groups.

One can also see this homomorphism explicitly on the standard complex, because we obtain a morphism of complexes

$$C(\lambda, \phi) : C(G, A) \to C(G', A')$$

which maps a continuous cochain f on the cochain $\varphi \circ f \circ \lambda^r$.

In particular, we have the inflation, lifting, restriction and conjugation:

$$\mathrm{inf} : H^r(G/N, A^N) \to H^r(G, A)$$
$$\mathrm{lif} : H^r(G/N, B) \to H^r(G, B)$$
$$\mathrm{res} : H^r(G, A) \to H^r(S, A)$$
$$\sigma_* : H^r(S, A) \to H^r(S^\sigma, \sigma^{-1}A),$$

for N closed normal in G, S closed in G and $\sigma \in G$.

132

All the commutativity relations of Chapter II are valid in the
present case, and we shall always refer to the corresponding result
in Chapter II when we want to apply the result to groups of Galois
type.

For U open but not necessarily normal in G, we also have the
transfer

$$\mathrm{tr} : H^r(U, A) \to H^r(G, A)$$

with $A \in \mathrm{Galm}(G)$. All the results of Chapter II, §1 for the transfer
also apply in the present case, because the proofs rely only on the
uniqueness theorem, the determination of the morphism in dimension 0, and the fact that injectives erase the cohomology functor in
dimension > 0.

(c) Limits. We have already seen in a naive way that our
cohomology functor on $\mathrm{Galm}(G)$ is a limit. We can state a more
general result as follows.

Theorem 2.4. *Let* (G_i, λ_{ij}) *and* (A_i, φ_{ij}) *be an inverse directed
family of groups of Galois type, and a directed system of abelian
groups respectively, on the same set of indices. Suppose that
for each* i, *we have* $A_i \in \mathrm{Galm}(G_i)$ *and that for* $i \leqq j$, *the
homomorphisms*

$$\lambda_{ij} : G_j \to G_i \quad and \quad \varphi_{ij} : A_i \to A_j$$

are compatible. Let $G = \mathrm{inv} \lim G_i$ *and* $A = \mathrm{dir} \lim A_i$. *Then* A
has a canonically determined structure as an element of $\mathrm{Galm}(G)$,
such that for each i, *the maps*

$$\lambda_i : G \to G_i \quad and \quad \varphi_i : A_i \to A$$

*are compatible. Furthermore, we have an isomorphism of com-
plexes*

$$\theta : C(G, A) \xrightarrow{\approx} \mathrm{dir} \lim C(G_i, A_i),$$

and consequently isomorphisms

$$\theta_* : H^r(G, A) \to \mathrm{dir} \lim H^r(G_i, A_i).$$

Proof. This is a generalization of the argument given for Theorem 2.3. It suffices to observe that each cochain $f : G^r \to A$ is
uniformly continuous, and consequently that there exists an open
normal subgroup U of G such that f depends only on cosets of U,

and takes on only a finite number of values. These values are all represented in some A_i. Hence there exists an open normal subgroup U_i of G_i such that $\lambda_1^{-1}(U_i) \subset U$, and we can construct a cochain $f_i : G_i^r \to A_i$ whose image in $C^r(G, A)$ is f. Similarly, we find that if the image of f_i in $C^r(G, A)$ is 0, then its image in $C^r(G_j, A_j)$ is also 0 for some $j > i, j$ sufficiently large. So the theorem follows.

We apply the preceding theorem in various cases, of which the most important are:

(a) When the G_i are all factor groups G/U with U open normal in G, the homomorphisms λ_{ij} then being surjective.

(b) When the G_i range over all open subgroups containing a closed subgroup S, the homomorphisms λ_{ij} then being inclusions.

Both cases are covered by the next lemma.

Lemma 2.5. *Let G be of Galois type, and let (G_i) be a family of closed subgroups, N_i a closed normal subgroup of G_i, indexed by a directed set $\{i\}$, and such that $N_j \subset N_i$ and $G_j \subset G_i$ when $i \leqq j$. Then one has*

$$\text{inv} \lim G_i/N_i = (\bigcap G_i)/(\bigcap N_i).$$

Proof. Clear.

Note that Theorem 2.3 is a special case of Theorem 2.4 (taking into account Lemma 2.5). In addition, we get more corollaries.

Corollary 2.6. *Let G be of Galois type and $A \in \text{Galm}(G)$. Let S be a closed subgroup of G. Then*

$$H^r(S, A) = \text{inv} \lim_V H^r(V, A),$$

the inverse limit being taken over all open subgroups V of G containing S.

Corollary 2.7. *Let $G, (G_i), (N_i)$ be as in Lemma 2.5. Let $A \in \text{Galm}(G)$ and let $N = \bigcap N_i$. Then*

$$H^r((\bigcap G_i)/(\bigcap N_i), A^N) \approx \text{dir} \lim H^r(G_i/N_i, A^{N_i}),$$

the limit being taken with respect to the canonical homomorphisms.

Proof. Immediate, because

$$A^N = \bigcup A^{N_i} = \text{dir lim } A^{N_i}$$

because by hypothesis $A \in \text{Galm}(G)$.

Corollary 2.8. *Let G be of Galois type and $A \in \text{Galm}(G)$. Then*

$$H^r(G, A) = \text{dir lim } H^r(G, E)$$

where the limit is taken with respect to the inclusion morphisms $E \subset A$, for all submodules E of A finitely generated over $\mathbf{Z}$.

Proof. By the definition of the continuous operation of G on A, we know that A is the union of G-submodules finitely generated over $\mathbf{Z}$, so we can apply the theorem.

Thus we see that the cohomology group $H^r(G, A)$ are limits of cohomology groups of finite groups, acting on finitely generated modules over $\mathbf{Z}$. We have already seen that these are torsion modules for $r > 0$.

Corollary 2.9. *Let m be an integer > 0, and $A \in \text{Galm}(G)$. Suppose*

$$m_A : A \to A$$

is an automorphism, in other words that A is uniquely divisible by m. Then the period of an element of $H^r(G, A)$ for $r > 0$ is an integer prime to m. If m_A is an automorphism for all positive integers m, then $H^r(G, A) = 0$ for all $r > 0$.

(d) The erasing functor, and induced representations.
We are going to define an erasing functor M_G on $\text{Galm}(G)$ similar to the one we defined on $\text{Mod}(G)$ when G is discrete.

Let S be a closed subgroup of G, which we suppose of Galois type. Let $B \in \text{Galm}(S)$ and let $M_G^S(B)$ be the set of all continuous maps $g : G \to B$ (B discrete) satisfying the relation

$$\sigma g(\tau) = g(\sigma\tau) \quad \text{for} \quad \sigma \in S, \tau \in G.$$

Addition is defined in $M_G^S(B)$ as usual, i.e. by adding values in B. We define an action of G by the formula

$$(\tau g)(x) = g(x\tau) \quad \text{for} \quad \tau, x \in G.$$

Because of the uniform continuity, one verifies at once that $M_G^S(B) \in \text{Galm}(G)$.

Taking into account the existence of a continuous section of G/S in G in Proposition 1.1, one sees that:

$M_G^S(B)$ is isomorphic to the Galois module of all continuous maps $G/S \to G$.

Thus we find results similar to those of Chapter II, which we summarize in a proposition.

Proposition 2.10. *Notations as above, M_G^S is a covariant, additive exact functor from $\text{Galm}(S)$ into $\text{Galm}(G)$. The bifunctors*

$$\text{Hom}_G(A, M_G^S(B)) \quad \text{and} \quad \text{Hom}_S(A, B)$$

on $\text{Galm}(G) \times \text{Galm}(S)$ are isomorphic. If B is injective in $\text{Galm}(S)$, then $M_G^S(B)$ is injective with $\text{Galm}(G)$.

The proof is the same as in Chapter II, in light of the condition of uniform continuity and the lemma on the existence of a cross section.

Theorem 2.11. *Let G be of Galois type, and S a closed subgroup. Then the inclusion $S \subset G$ is compatible with the homomorphism*

$$g \mapsto g(e) \quad \text{of} \quad M_G^S(B) \to B,$$

giving rise to an isomorphism of functors

$$H_G \circ M_G^S \approx H_S.$$

In particular, if $S = e$, then $H^r(G, M_S(B)) = 0$ for $r > 0$.

Proof. Identical to the proof when G is discrete. For the last assertion, when $S = e$, we put $M_G = M_G^e$.

In particular, we obtain an erasing functor $M_G = M_G^e$ as in the discrete case. For $A \in \text{Galm}(G)$, we have an exact sequence

$$0 \to A \xrightarrow{\varepsilon_A} M_G(A) \to X(A) \to 0,$$

where ε_A is defined by the formula $\varepsilon_A(a) = g_a$ and $g_a(\sigma) = \sigma a$ for $\sigma \in G$.

As in the discrete case, the above exact sequence splits.

Corollary 2.12. *Let G be of Galois type, S a closed subgroup, and $B \in \mathrm{Galm}(S)$. Then $H^r(S, M_G^S(B)) = 0$ for $r > 0$.*

Proof. When $S = G$, this is a special case of the theorem, taking $S = e$. If V ranges over the family of open subgroups containing S, then we use the fact of Corollary 2.6 that

$$H^r(S, A) = \text{dir lim } H^r(V, A).$$

It will therefore suffice to prove the result when $S = V$ is open. But in this case, $M_S(B)$ is isomorphic in $\mathrm{Galm}(V)$ to a finite product of $M_G^V(B)$, and one can apply the preceding result.

Corollary 2.13. *Let $A \in \mathrm{Galm}(G)$ be injective. Then $H^r(S, A) = 0$ for all closed subgroups S of G and $r > 0$.*

Proof. In the erasing sequence with ε_A, we see that A is a direct factor of $M_G(A)$, so we can apply Corollary 2.12.

(e) Cup products. The theory of cup products can be developed exactly as in the case when G is discrete. Since existence was proved previously with the standard complex, using general theorems on abelian categories, we can do the same thing in the present case. In addition, we observe that

$\mathrm{Galm}(G)$ *is closed under taking the tensor product,*

as one sees immediately, so that tensor products can be used to factorize multilinear maps. Thus $\mathrm{Galm}(G)$ can be defined to be a multilinear category. If $A_1, \ldots, A_n, B$ are in $\mathrm{Galm}(G)$, then we define $f : A_1 \times \ldots \times A_N \to B$ to be in $L(A_1, \ldots, A_n, B)$ if f is multilinear in $\mathrm{Mod}(\mathbf{Z})$, and

$$f(\sigma a_1, \ldots, \sigma a_n) = \sigma f(a_1, \ldots, a_n) \quad \text{for all} \quad \sigma \in G,$$

exactly as in the case where G is discrete.

We thus obtain the existence and uniqueness of the cup product, which satisfies the property of the three exact sequences as in the discrete case. Again, we have the same relations of commutativity concerning the transfer, restriction, inflation and conjugation.

(f) Spectral sequence. The results concerning spectral sequences apply without change, taking into account the uniform continuity of cochains. We have a functor $F : \mathrm{Galm}(G) \to \mathrm{Galm}(G/N)$ for a closed normal subgroup N, defined by $A \mapsto A^N$. The group of Galois type G/N acts on $H^r(N, A)$ by conjugation, and one has:

Proposition 2.14. *If N is closed normal in G, then $H^r(N, A)$ is in $\mathrm{Galm}(G/N)$ for $A \in \mathrm{Galm}(G)$.*

Proof. If $\sigma \in N$, from the definition of σ_*, we know that $\sigma_* = \mathrm{id}$. We have to show that for all $\sigma \in H^r(N, A)$ there exists an open subgroup U such that $\sigma_*\alpha = \alpha$ for all $\sigma \in U$. But by shifting dimensions, there exist exact sequences and coboundaries $\delta_1, \ldots \delta_r$ such that

$$\alpha = \delta_1, \ldots, \delta_r \alpha_0 \quad \text{with } \alpha_0 \in H^0(N, B) \text{ for some } B \in \mathrm{Galm}(G).$$

One merely uses the erasing functor r times. We have

$$\sigma_*\alpha = \sigma_* \delta_1 \ldots \delta_r \alpha_0 = \delta_1 \ldots \delta_r \sigma_* \alpha_0$$

and we apply the result in dimension 0, which is clear in this case since σ_* denotes the continuous operation of $\sigma \in S$.

Since the functor $A \mapsto A^N$ transforms an injective module to an injective module, one obtains the spectral sequence of the composite of derived functors. The explicit computations for the restriction, inflation and the edge homomorphisms remain valid in the present case.

(g) Sylow subgroups. As a further application of the fact that the cohomology of Galois type groups is a limit of cohomology of finite groups, we find:

Proposition 2.15. *Let G be of Galois type, and $A \in \mathrm{Galm}(G)$. Let S be a closed subgroup of G. If $(G : S)$ is prime to a prime number p, then the restriction*

$$\mathrm{res} : H^r(G, A) \to H^r(S, A)$$

induces an injection on $H^r(G, A, p)$.

Proof. If S is an open subgroup V in G, then we have the transfer and restriction formula

$$\mathrm{tr} \circ \mathrm{res}(\alpha) = (G : V)\alpha,$$

which proves our assertion. The general case follows, taking into account that

$$H^r(S, A) = \text{dir lim } H^r(V, A)$$

for V open containing S.

§3. Cohomological dimension.

Let G be a group of Galois type. We denote by $\text{Galm}_{\text{tor}}(G)$ the abelian category whose objects are the objects A of $\text{Galm}(G)$ which are torsion modules, i.e. for each $a \in A$ there is an integer $n \neq 0$ such that $na = 0$. Given $A \in \text{Galm}(G)$, we denote by A_{tor} the submodule of torsion elements. Similarly for a prime p, we let A_{p^n} denote the kernel of p_A^n in A, and A_{p^∞} is the union of all A_{p^n} for all positive integers n. We call A_{p^∞} the submodule of **p-primary elements**. As usual for an integer m, we let A_m be the kernel of m_A, so

$$A_{\text{tor}} = \bigcup A_m \quad \text{and} \quad A_{p^\infty} = \bigcup A_{p^n}$$

the first union being taken for $m \in \mathbf{Z}, m > 0$ and the second for $n > 0$.

The subcategory of elements $A \in \text{Galm}(G)$ such that $A = A_{p^\infty}$ (i.e. A is p-primary) will be denoted by $\text{Galm}_p(G)$.

Let n be an integer > 0. We define the notion of **cohomological dimension**, abbreviated **cd**, and **strict cohomological dimension**, abbreviated **scd**, as follows.

$\text{cd}(G) \leqq n$ if and only if $H^r(G, A) = 0$ for all $r > n$
         and $A \in \text{Galm}_{\text{tor}}(G)$

$\text{cd}_p(G) \leqq n$ if and only if $H^r(G, A, p) = 0$ for all $r > n$
         and $A \in \text{Galm}_{\text{tor}}(G)$

$\text{scd}(G) \leqq n$ if and only if $H^r(G, A) = 0$ for all $r > n$
         and $A \in \text{Galm}(G)$

$\text{scd}_p(G) \leqq n$ if and only if $H^r(G, A, p) = 0$ for $r > n$
         and $A \in \text{Galm}(G)$.

We note that cohomological dimension is defined via torsion modules, and the strict cohomological dimension is defined by means of arbitrary modules (in $\text{Galm}(G)$, of course).

Since

$$H^r(G, A) = \bigoplus_p H^r(G, A, p)$$

one sees that

$$\operatorname{cd}(G) = \sup_p \operatorname{cd}_p(G) \quad \text{and} \quad \operatorname{scd}(G) = \sup_p \operatorname{scd}_p(G).$$

For all $A \in \operatorname{Galm}_{\operatorname{tor}}(G)$ we have $A = \bigcup A_{p^\infty}$, the direct sum being taken over all primes p. Hence

$$H^r(G, A) = \bigoplus H^r(G, A_{p^\infty}).$$

To determine $\operatorname{cd}_p(G)$, it will suffice to consider $H^r(G, A_{p^\infty})$, because if we let $A'_{(p)}$ be the p-complementary module

$$A'_{(p)} = \bigcup A_m \quad \text{with} \quad m \quad \text{prime to} \quad p,$$

then $A'_{(p)}$ is uniquely determined by p^n for all integers $n > 0$, so p^n induces an automorphism of $H^r(G, A'_{(p)})$ for $r > 0$, and $H^r(G, A'_{(p)})$ is a torsion group. Hence $H^r(G, A'_{(p)})$ does not contain any element whose torsion is a power of p, and we find:

Proposition 3.1. *Let $A \in \operatorname{Galm}_{\operatorname{tor}}(G)$. Then the homomorphism*

$$H^r(G, A_{p^\infty}) \to H^r(G, A, p)$$

induced by the inclusion $A_{p^\infty} \subset A$ is an isomorphism for all r.

Corollary 3.2. *In the definition of $\operatorname{cd}_p(G)$, one can replace the condition $A \in \operatorname{Galm}_{\operatorname{tor}}(G)$ by $A \in \operatorname{Galm}_p(G)$.*

We are going to see that the strict dimension can differ only by 1 from the other dimension.

Proposition 3.3. *Let G be of Galois type, and p prime. Then*

$$\operatorname{cd}_p(G) \leqq \operatorname{scd}_p(G) \leqq \operatorname{cd}_p(G) + 1,$$

and the same inequalities hold omitting the index p.

Proof. The first inequality is trivial. For the second, consider the exact sequence

$$0 \to pA \xrightarrow{i} A \to A/pA \to 0$$

$$0 \to A_p \to A \xrightarrow{j} pA \to 0$$

and the corresponding cohomology exact sequences

$$H^{r+1}(pA) \xrightarrow{i_*} H^{r+1}(A) \to H^{r+1}(A/pA)$$

$$H^{r+1}(A_p) \to H^{r+1}(A) \xrightarrow{j_*} H^{r+1}(A/pA).$$

We assume that $cd_p(G) < n$ and $r > n$. Since $ij = p$, we find $i_* j_* = p_*$. We have $H^{r+1}(A_p) = 0$ by definition, and also $H^{r+1}(A/pA) = 0$. One then sees that j_* is bijective and i_* is surjective. Hence p_* is surjective, i.e. $H^r(A)$ is divisible by p, and hence by an arbitrary power of p. The elements of $H^r(G, A, p)$ being p-primary, it follows that $H^{r+1}(G, A, p) = 0$. This proves the proposition.

For the next result, we need a lemma on the erasing functor M_G^S.

Lemma 3.4. *Let G be of Galois type, and S a closed subgroup. Let $B \in \mathrm{Galm}_{tor}(S)$ (resp. $\mathrm{Galm}_p(S)$). Then $M_G^S(B)$ is in $\mathrm{Galm}_{tor}(S)$ (resp. $\mathrm{Galm}_p(G)$). If, in addition B is finitely generated over $\mathbf{Z}$, and S is open, then $M_G^S(B)$ is finitely generated over $\mathbf{Z}$.*

Proof. Immediate from the definitions.

Proposition 3.5. *Let S be a closed subgroup of H. Then*

$$cd_p \leqq cd_p(G) \quad and \quad scd_p(S) \leqq scd_p(G),$$

and equality holds if $(G : S)$ is prime to p.

Proof. By Theorem 2.11, we know that $H^r(G, M_G^S(B)) \approx H^r(S, B)$ for all $B \in \mathrm{Galm}(S)$. The assertions are then immediate consequences of the definitions, together with the fact that $(G : S)$ prime to p implies that the restriction is an injection on the p-primary part of cohomology (Proposition 2.15).

As a special case, we find:

Corollary 3.6. *Let G_p be a p-Sylow subgroup of G. Then*

$$\text{cd}_p(G) = \text{cd}_p(G_p) = \text{cd}(G_p),$$

and similarly with scd *instead of* cd. *Furthermore,*

$$\text{cd}(G) = \sup_p \text{cd}(G_p) \quad and \quad \text{scd}(G) = \sup_p \text{scd}(G_p).$$

We now study the cohomological dimension, and leave aside the strict dimension. First, we have a criterion in terms of a category of submodules, easily described.

Proposition 3.7. *We have* $\text{cd}_p(G) \leqq n$ *if and only if* $H^{n+1}(G, E) = 0$ *for all elements* $E \in \text{Galm}(G)$ *such that E is finite of p-power order, and simple as a G-module.*

Proof. Implication in one direction is trivial, taking into account that E is uniquely divisible by every integer m prime to p, and therefore that

$$H^{n+1}(G, E) = H^{n+1}(G, E, p).$$

Conversely, suppose $H^{n+1}(G, E) = 0$ for all E as prescribed. Let $A \in \text{Galm}_{\text{tor}}(G)$ have finite p-power order. If $A \neq 0$, then there is an exact sequence

$$0 \to A' \to A \to A'' \to 0$$

with A' simple. The order of A'' is strictly smaller then the order of A, and the exact cohomology sequence shows by induction that $H^{n+1}(G, A) = 0$. Then let $A \in \text{Galm}_p(G)$. Then A is a direct limit of finite submodules, and we can apply Corollary 2.8. It follows that $H^{n+1}(G, A) = 0$ for $A \in \text{Galm}_p(G)$. Using the erasing functor M_G, one can then proceed by induction, taking into account the fact that M_G maps $\text{Galm}_p(G)$ into $\text{Galm}_p(G)$, and one finds $H^r(G, A) = 0$ for $r > n$ and $A \in \text{Galm}_p(G)$. We can conclude the proof by applying Corollary 3.2.

Lemma 3.8. *Let G be a p-group of Galois type, and let* $A \in \text{Galm}_p(G)$. *If $A^G = 0$ then $A = 0$. The only simple module* $A \in \text{Galm}_p(G)$ *is* $\mathbf{F}_p$.

Proof. We already proved this lemma when G is finite, and the general case is an immediate consequence, because G acts continuously on A.

Theorem 3.9. *Let $G = G_p$ be a p-group of Galois type. Then* $\mathrm{cd}(G) \leqq n$ *if and only if* $H^{n+1}(G, \mathbf{F}_p) = 0$.

Proof. This is immediate from Proposition 3.7 and the lemma.

Theorem 3.10. *Let G be a group of Galois type. The following conditions are equivalent:*

$\mathrm{cd}(G) = 0$;
$\mathrm{scd}(G) = 0$;
$G = e$.

If G is a p-group, then $\mathrm{cd}_p(G) = 0$ *implies* $G = e$.

Proof. It will clearly suffice to prove that if $\mathrm{cd}(G) = 0$ then G is trivial, so suppose $\mathrm{cd}(G) = 0$. For every p-Sylow subgroup G_p of G, we have $\mathrm{cd}(G_p) = 0$ (as one sees from the induced representation), and

$$\mathrm{cd}(G_p) = \mathrm{cd}_p(G_p).$$

Hence $H^1(G_p, \mathbf{F}_p) = 0$. But G_p acts trivially on $\mathbf{F}_p$ so $H^1(G_p, \mathbf{F}_p)$ is just the group of continuous homomorphism cont $\hom(G_p, \mathbf{F}_p)$. If $G \neq e$, then there exists an open normal subgroup U such that G/U is a finite p-group, is equal to e. One could then construct a non-trivial homomorphism of G/U into $\mathbf{F}_p$, contradicting the hypothesis, and concluding the proof.

To show that certain cohomology groups are not 0 in certain dimensions greater than some integer, we have the following criterion.

Lemma 3.11. *Let G be a p-group of Galois type and* $\mathrm{cd}(G) = n < \infty$. *If $E \in \mathrm{Galm}_p(G)$ has finite order and $E \neq 0$, then* $H^n(G, E) \neq 0$.

Proof. By Lemma 3.8, there is an exact sequence

$$0 \to E' \to E \to E'' \to 0$$

with a maximal submodule E' of E. Since $H^n(G, \mathbf{F}_p) \neq 0$ by hypothesis, one has, again by hypothesis, the exact sequence

$$H^n(G, E) \to H^n(G, \mathbf{F}_p) \to H^{n+1}(G, E') = 0,$$

which shows that $H^n(G, E)$ cannot be trivial.

As an application, we give a refinement of Proposition 3.5.

Proposition 3.12. *Let G be of Galois type, and let S be a closed subgroup of G. If $\mathrm{ord}_p(G:S)$ is finite and $\mathrm{cd}_p(G) < \infty$, then $\mathrm{cd}_p(S) = \mathrm{cd}_p(G)$.*

Proof. Let S_p be a p-Sylow subgroup of S, and similarly G_p a p-Sylow subgroup of G containing S_p. Then

$$\mathrm{ord}_p(G_p : S_p) + \mathrm{ord}_p(G : G_p) = \mathrm{ord}_p(G : S_p)$$
$$= \mathrm{ord}_p(G : S) + \mathrm{ord}_p(S : S_p)$$

Hence $\mathrm{ord}_p(G_p : S_p) = \mathrm{ord}_p(G : S)$. This reduces the proof to the case when G is a p-group, and S is open in G. Suppose

$$n = \mathrm{cd}(G) < \infty.$$

Then by Lemma 3.11,

$$H^n(S, \mathbf{F}_p) = H^n(G, M_G^S(\mathbf{F}_p)) \neq 0,$$

because $M_G^S(\mathbf{F}_p)$ has $p^{(G:S)}$ elements. This concludes the proof.

Corollary 3.13. *If $0 < \mathrm{ord}_p(G : e) < \infty$, then $\mathrm{cd}_p(G) = \infty$. In fact, if G is a finite p-group, then $H^r(G, \mathbf{F}_p) \neq 0$ for all $r > 0$.*

From this corollary, one sees the cohomological dimension is interesting only for infinite groups. We shall give below examples of Galois groups with finite cohomological dimension.

§4. Cohomological dimension $\leqq 1$.

Let us first remark that if G is a group of Galois type with $\mathrm{scd}_p(G) \leqq 1$, then $\mathrm{scd}_p(G) = 0$ and hence every p-Sylow subgroup G_p of G is trivial. Indeed, we have by hypothesis

$$0 = H^2(G_p, \mathbf{Z}) \approx H^1(G_p, \mathbf{Q}/\mathbf{Z}) = \mathrm{cont}\ \mathrm{hom}(G_p, \mathbf{Q}/\mathbf{Z})$$

from the exact sequence with $\mathbf{Z}, \mathbf{Q}$ and $\mathbf{Q}/\mathbf{Z}$. That $\mathbf{Q}$ is uniquely divisible by every integer $\neq 0$ implies that its cohomology is 0 in dimensions > 0. At the end of the preceding section, we saw that if $G_p \neq e$ then we can find a non-trivial continuous homomorphism of G_p into $\mathbf{F}_p$, which can be naturally imbedded in $\mathbf{Q}/\mathbf{Z}$, and one sees therefore that $G_p = e$, thus proving our assertion.

We then consider the condition $\mathrm{cd}_p(G) \leqq 1$. We shall see that this condition characterizes certain topologically free groups.

We define a group of Galois type to be *p-extensive* if and only if for every finite group F and each abelian p-subgroup E normal in F, and every continuous homomorphism $f : G \to F/E$, there exists a continuous homomorphism $\bar{f} : G \to F$ which makes the following diagram commutative:

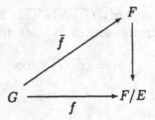

Proposition 4.1. *We have* $\mathrm{cd}_p(G) \leqq 1$ *if and only if G is p-extensive.*

Proof. Suppose first that $\mathrm{cd}_p(G) \leqq 1$. We are given F, E, f as above. As usual, we may consider E as an F/E-module, the operation being that of conjugation. Consequently, E is in $\mathrm{Galm}(G)$ via f, namely for $\sigma \in G$ and $x \in E$ we define

$$\sigma x = f(\sigma) x.$$

For each $\sigma \in F/E$, let u_σ be a representative in F. Put

$$e_{\sigma,\tau} = u_\sigma u_\tau u_{\sigma\tau}^{-1}.$$

Then $(e_{\sigma,\tau})$ is a 2-cocycle in $C^2(F/E, E)$, and consequently $(e_{f(\sigma),f(\tau)})$ is a 2-cocycle in $C^2(G, E)$. By hypothesis, there exists a continuous map $\sigma \mapsto a_\sigma$ of G in E such that

$$e_{f(\sigma),f(\tau)} = a_{\sigma\tau}/a_\sigma \sigma a_\tau.$$

We define $\bar{f}(\sigma)$ by

$$\bar{f}(\sigma) = a_\sigma u_{f(\sigma)}.$$

From the definition of the action of G on E, we have

$$\sigma a = u_{f(\sigma)} a u_{f(\sigma)}^{-1}.$$

Thus we find

$$\bar{f}(\sigma)\bar{f}(\sigma) = a_\sigma u_{f(\sigma)} a_\tau u_{f(\tau)} = a_\sigma a_\tau^\sigma u_{f(\sigma)} u_{f(\tau)}$$
$$= a_\sigma a_\tau^\sigma e_{f(\sigma),f(\tau)} u_{f(\sigma\tau)}$$
$$= a_{\sigma\tau} u_{f(\sigma\tau)}$$
$$= \bar{f}(\sigma\tau),$$

which shows that $\bar{f}$ is a homomorphism. It is continuous because (a_σ) is a continuous cochain, and $\sigma \mapsto f(\sigma)$ is continuous. Furthermore, it is clear that $\bar{f}$ is a lifting of f, i.e. that the diagram as in the definition of p-extensive is commutative.

Conversely, let $E \in \mathrm{Galm}_{\mathrm{tor}}(G)$ be of finite order, equal to a p-power, and let $\alpha \in H^2(G, E)$. We have to prove that $\alpha = 0$. Since E is finite, there exists an open normal subgroup U such that U leaves E fixed, i.e. $E = E^U$, and E is therefore a G/U-module. Taking a smaller open subgroup of U if necessary, we can suppose without loss of generality that α comes from the inflation of an element in $H^2(G/U, E)$, i.e. there exists $\alpha_0 \in H^2(G/U, E)$ such that $\alpha = \inf_G^{G/U}(\alpha_0)$. Let F be the group extension of G/U by E corresponding to the class of α_0, so that we have $G/U = F/E$, and let

$$f : G \to F/E$$

be the corresponding homomorphism. We are then in the same situation as in the first part of the proof, and $(e_{f(\sigma),f(\tau)})$ is a 2-cocycle representing α. Since $\bar{f}$ now exists by hypothesis, we define $a_\sigma = \bar{f}(\sigma) u_{f(\sigma)}^{-1}$. The same computation as before shows that

$$a_{\sigma,\tau} = a_\sigma a_\tau^\sigma e_{f(\sigma),f(\tau)},$$

and since (a_σ) is clearly a continuous cochain, one sees that $(e_{f(\sigma),f(\tau)})$ is a coboundary, in other words $\alpha = 0$. This concludes the proof.

Remark. In the definition of p-extensive, without loss of generality, we may assume that f is surjective (it suffices to replace F

by the inverse image of $f(G)$ in F/E). However, we cannot require that $\bar{f}$ is surjective. For instance, let G be the Galois group of the separable closure of a field k. Then F/E is the Galois group of a finite extension K/k, and the problem of finding $\bar{f}$ surjective amounts to finding a finite Galois extension $L \supset K \supset k$ such that F is its Galois group, a problem considered for example by Iwasawa, *Annls of Math.* 1953.

We shall now extend the extension property to the situation when we can take F, E to be of Galois type.

Proposition 4.2. *Let G be of Galois type and p-extensive. Then the p-extension property concerning $(G, f, F, F/E)$ is valid when F is of Galois type (rather than finite), and E is a closed normal p-subgroup.*

Proof. We suppose first that E is finite abelian normal in F. There exists an open normal subgroup U such that $U \cap E = e$. Let $f_1 : G \to F/EU$ be the composite of $f : G \to F/E$ with the canonical homomorphism $F/E \to F/EU$. We can lift f_1 to a continuous homomorphism $\bar{f}_1 : G \to F/U$ by p-extensivity for $F_1 = F/U$ and $E_1 = EU/(U \cap E)$, and f_1. We have a homomorphism

$$(f, \bar{f}_1) : G \to (F/E) \times (F/U),$$

and the canonical map $i : F \to (F/E) \times (F/U)$ is an injection since $U \cap E = e$. The image of G under $(f, \bar{f}_1)$ is contained in the image of i because f and $\bar{f}_1$ lift f_1. Hence $\bar{f} = (f, \bar{f}_1) : G \to F$ solves the extension problem in the present case.

We can now deal with the general case. We want to lift $f : G \to F/E$. We consider all pairs (E', f') where E' is a closed subgroup of E normal in F, and $f' : G \to F/E'$ lifts f. By Zorn's lemma, there is a maximal pair, which we denote also by (E, f). We have to show that $E = e$. If $E \neq e$, then there exists a non trivial element $\theta \in H^1(E, \mathbf{F}_p)$. This character vanishes on an open subgroup V, and has therefore only a finite number of conjugates by elements of F, i.e. it is a Galois module of F/E. Let E_1 be the intersection of the kernel of θ and all its conjugates. Then E_1 is a closed subgroup of E, normal in F, and by the first part of the proof we can lift f to $f_1 : G \to F/E_1$, which contradicts the hypothesis that (E, f) is maximal, and concludes the proof of Proposition 4.2.

Next, we connect cohomological dimension with free groups. We fix a prime p.

Let X be a set and $F_0(X)$ the free group generated by X in the ordinary meaning of the word (cf. *Algebra*, Chapter I, §12). We consider the family of normal subgroups $U \subset X$ such that:

(i) U contains all but a finite number of elements of X.

(ii) U has index a power of p in $F_0(X)$.

We let $F_p(X)$ be the inverse limit

$$F_p(X) = \text{inv lim } F_0/U$$

taken over all such subgroups U. We call $F_p(X)$ the **profinite free p-group generated by** X. Thus $F_p(X)$ is a group of Galois type.

Let G be a group of Galois type and G^0 the intersection of all the kernels of continuous homomorphisms $\theta : G \to \mathbf{F}_p$, i.e. $\theta \in H^1(G, \mathbf{F}_p)$. Then $H^1(G, \mathbf{F}_p)$ is the character group of G/G^0. The converse is also true by Pontrjagin duality.

By definition, if P is a finite p-group, then the continuous homomorphisms $f : F_p(X) \to P$ are in bijection with the maps $f_0 : X \to P$ such that $f_0(x) = e$ for all but a finite number of $x \in X$. Hence $H^1(F_p(X), \mathbf{F}_p)$ is a vector space over $\mathbf{F}_p$, of finite dimension equal to the cardinality of X, and having a basis which can be identified with the elements of X.

Furthermore, we see that $F_p(X)$ is p-extensive, and that cd $F_p(X) \leqq 1$. Indeed, we can take f surjective in the definition of p-extensive, and F is then a finite p-group. One uses the freeness of F_0 to see immediately that $F_p(X)$ is p-extensive. We shall prove the converse.

Lemma 4.3. *Let G be a p-group of Galois type, S a closed subgroup. Then $SG^0 = G$ implies $S = G$.*

Proof. This is essentially an analogue of Nakayama's lemma in commutative algebra. Actually, one can prove the lemma first for finite groups, and then extend it to the infinite case, to be left to the reader.

Theorem 4.4. *Let G be a p-group of Galois type. Then there exists a profinite free p-group, $F_p(X)$ and a continuous homomorphism*

$$\bar{g} : F_p(X) \to G$$

such that the induced homomorphism

$$H^1(G, \mathbf{F}_p) \to H^1(F_p(X), \mathbf{F}_p)$$

is an isomorphism. The map $\bar{g}$ is then surjective. If $\mathrm{cd}(G) \leqq 1$, then $\bar{g}$ is an isomorphism.

Proof. From the preceding discussion, to obtain an isomorphism

$$H^1(G, \mathbf{F}_p) \xrightarrow{\approx} H^1(F_p(X), \mathbf{F}_p)$$

it suffices to take for X a basis of $H^1(G, \mathbf{F}_p)$ and to form $F_p(X)$. By duality, we obtain an isomorphism

$$F_p(X)/F_p(X)^0 \approx G/G^0,$$

whence a homomorphism

$$g : F_p(X) \to G/G^0.$$

Since $F_p(X)$ is p-extensive, we can lift g to G, to get the commutative diagram

and $\bar{g}$ is surjective by the lemma. Suppose finally that $\mathrm{cd}(G) \leqq 1$, and let N be the kernel of $\bar{g}$. Then we obtain an exact sequence

$$0 \to H^1(G,\mathbf{F}_p) \xrightarrow{\text{inf}} H^1(F_p(X),\mathbf{F}_p) \xrightarrow{\text{res}} H^1(N,\mathbf{F}_p)^G \to H^2(G,\mathbf{F}_p)=0.$$

We have $H^2(G, \mathbf{F}_p) = 0$ by the assumption $\mathrm{cd}(G) \leqq 1$. The inflation is an isomorphism, and hence $H^1(N, \mathbf{F}_p)^G = 0$. By Lemma 3.8, we find $H^1(N, \mathbf{F}_p) = 0$, i.e. N has only the trivial character, whence $N = e$, thus proving the theorem.

Corollary 4.5. *Let G be a p-group of Galois type. Then the following conditions are equivalent:*

G is profinite free;

G is p-extensive;

$\mathrm{cd}(G) \leqq 1$.

We end this section with a discussion of the condition $\mathrm{cd}(G) \leqq 1$ for factor groups. Let G be of Galois type. Let T be the intersection of all subgroups of G which are the kernels of continuous homomorphisms of G into p-groups of Galois type. Then G/T is a p-group which we denote by $G(p)$, and which we call the **maximal p-quotient** of G. One can also characterize T by the condition that it is a closed normal subgroup satisfying:

(a) $(G : T)$ is a p-power.

(b) $H^1(T, \mathbf{F}_p) = 0$.

The characterization is immediate.

Proposition 4.6. *Let G be a group of Galois type. Then* $\mathrm{cd}_p(G) \leqq 1$ *implies* $\mathrm{cd}_p G(p) \leqq 1$.

Proof. Consider the exact sequence

$$0 \rightarrow H^1(G/T, \mathbf{F}_p) \rightarrow H^1(G, \mathbf{F}_p) \rightarrow H^1(T, \mathbf{F}_p)^{G/T} \rightarrow H^2(G/T, \mathbf{F}_p) \rightarrow 0,$$

with a 0 on the right because of the assumption $\mathrm{cd}_p(G) \leqq 1$. By the characterization of T we have $H^1(T, \mathbf{F}_p) = 0$, whence $H^2(G/T, \mathbf{F}_p) = 0$ which suffices to prove the proposition by Theorem 3.9.

In the Galois theory, $G(p)$ is the Galois group of the maximal p-extension of the ground field, and G is the Galois group of the algebraic closure. Cf. §6 below for applications to this context.

§5. The tower theorem

In many cases, one gets information on a group G be considering a normal subgroup N and the factor group G/N. We do this for cohomological dimension, and we shall find

$$\mathrm{cd}(G) \leqq \mathrm{cd}(N) + \mathrm{cd}(G/N),$$

and similar with cd_p instead of cd. We use the spectral sequence with

$$E_2^{r,s} = H^r(G/N, H^s(N, A)) \quad \text{converging to} \quad H(G, A)$$

for $A \in \mathrm{Galm}(G)$. There is a filtration of $H^n(G, A)$ such that the successive quotients are isomorphic to $E_\infty^{r,s}$ for $r + s = n$, and $E_\infty^{r,s}$ is a subgroup of a factor group of $E_2^{r,s}$. Hence $H^n(G, A) = 0$ whenever $H^r(G/N, H^s(N, A)) = 0$, which occurs in the following cases:

$$r > \mathrm{cd}(G/N) \quad \text{and} \quad s > 0 \text{ or } A \in \mathrm{Galm}_{\mathrm{tor}}(G);$$
$$r > \mathrm{scd}(G/N) \quad \text{and} \quad s \text{ arbitrary};$$
$$s > \mathrm{cd}(N) \quad \text{and} \quad A \in \mathrm{Galm}_{\mathrm{tor}}(G);$$
$$s > \mathrm{scd}(N) \quad \text{and} \quad A \in \mathrm{Galm}(G).$$

From this we find the theorem:

Theorem 5.1. *Let G be of Galois type and N a closed normal subgroup. Then for all primes p,*

$$\mathrm{cd}_p(G) \leqq \mathrm{cd}_p(G/N) + \mathrm{cd}_p(N),$$

and similarly with cd instead of cd_p.

As an application, suppose that G/N is topologically cyclic, and $\mathrm{cd}_p(N) \leqq 1$. Then $\mathrm{cd}_p(G) \leqq 2$. This happens in the following cases: G is the Galois group of the algebraic closure of a totally imaginary number field, or a p-adic field. Indeed, in each case, one can construct a cyclic extension (maximal unramified in the local case, cyclotomic in the global case), which decomposes G into a subgroup N and factor G/N as above. In the next sections, we shall give a criterion with the Brauer group to show that $\mathrm{cd}(N) \leqq 1$ for suitable N.

§6. Galois groups over a field

Let k be a field and k_s its separable closure. Let

$$G_k = \mathrm{Gal}(K_s/k)$$

be the Galois group. If K is a Galois extension of k, we let $G_{K/k}$ be its Galois group. Then G_K is normal in G_k and the factor group G_k/G_K is $G_{K/k}$. All these groups are of Galois type, with the Krull topology.

We shall use constantly Hilbert's Theorem 90, that for the multiplicative group K^*, we have

$$H^1(G_{K/k}, K^*) = 0.$$

Note that $K^* \in \mathrm{Galm}(G_{K/k})$. In the additive case, with the additive group K^+,

$$H^r(G_{K/k}, K^+) = 0 \quad \text{for all} \quad r > 0.$$

One sees this reduction to the case when K is finite Galois over k, so there is a normal basis showing that K^+ is semilocal with local group reduced to e, whence the cohomology is trivial in dimension > 0.

Next, we give a result in characteristic p.

Theorem 6.1. *Let k have characteristic $p > 0$ and let $k(p)$ be the maximal p-extension with Galois group $G(p)$ over k. Then cd $G(p) \leq 1$, and so $G(p)$ is profinite free. The number of generators is equal to $\dim_{F_p}(k^+/\wp k^+)$, where $\wp x = x^p - x$.*

Proof. We recall the Kummer theory exact sequence

$$0 \to F_p \to k_s^+ \xrightarrow{\wp} k_s^+ \to 0,$$

whence the cohomology sequence

$$0 \to \mathbf{F}_p \to k^+ \to k^+ \to H^1(G_k, \mathbf{F}_p) \to H^1(G_k, k_s^+) = 0.$$

Consequently,

$$k^+/\wp k^+ \approx H^1(G_k, \mathbf{F}_p) = \mathrm{cont\ hom}(G_k, \mathbf{F}_p).$$

Since $k(p)$ is the maximal Galois p-extension of k, it has no Galois extension of p-power degree, and hence we have an exact sequence

$$0 \to \mathbf{F}_p \to k(p)^+ \to k(p)^+ \to 0.$$

By the remarks made at the beginning of this section, we get from the exact cohomology sequence that $H^2(G(p), \mathbf{F}_p) = 0$, and hence by the criterion of Theorem 3.9 that cd $G(p) \leqq 1$. Moreover, the beginning of this same exact sequence yields

$$k^+ \xrightarrow{p} k^+ \to H^1(G(p), \mathbf{F}_p) \to H^1(G(p), k(p)^+) = 0,$$

whence an isomorphism

$$k^+/\wp k^+ \approx H^1(G(p), \mathbf{F}_p)),$$

which gives us the desired number of generators.

We now go to characteristic $\neq p$ using the multiplicative Kummer sequence instead of the additive one. Cohomological dimension will be studied via Galois cohomology in k_s^*.

Theorem 6.2. *Let k be a field of characteristic $\neq p$ and containing a p-th root of unity. Let $k(p)$ be the maximal p-extension, with Galois group $G(p)$. Then $\mathrm{cd}(G(p)) \leqq n$ if and only if:*

(i) *$H^n(G(p), k(p)^*)$ is divisible by p.*

(ii) *$H^{n+1}(G(p), k(p)^*) = 0$.*

Proof. We consider the exact sequence

$$0 \to \mathbf{F}_p \to k(p)^* \xrightarrow{p} k(p)^* \to 0,$$

where $\mathbf{F}_p$ gets embedded on the group of p-th roots of unity, and the map on the right is taking p-th powers. We obtain the cohomology exact sequence

$$H^n(k(p)^*) \to H^n(k(p)^*) \to H^{n+1}(\mathbf{F}_p) \to H^{n+1}(k(p)^*) \to H^{n+1}(k(p)^*)$$

with acting group $G(p)$. The theorem follows at once from this exact sequence.

Corollary 6.3. (Kawada) *The Galois group $G(p) = G_{k(p)/k}$ is a profinite p-group if $p = \mathrm{char}\ k$. If $p \neq \mathrm{char}\ k$ and the p-th roots of unity are in k, then it is profinite free if and only if $H^2(G(p), k(p)) = 0$.*

Proof. One always has $H^1(k(p)^*) = 0$, and the rest follows from the preceding two theorems.

Theorem 6.1 can be translated in terms of cohomology with values in k_s^*.

Theorem 6.4. *Let k be a field and p prime $\neq$ char k. Let n be an integer > 0. Then $\mathrm{cd}_p(G_k) \leqq n$ if and only if:*

(i) $H^n(G_E, k_s^*) = 0$ *is divisible by* p

(ii) $H^{n+1}(G_E, k_s^*, p) = 0$

for all finite separable extensions E of k of degree prime to p.

Proof. The Kummer sequence

$$0 \to \mathbf{F}_p \to k_s^* \xrightarrow{p} k_s^* \to 0$$

yields the cohomology sequence with groups G_k:

$$H^n(k_s^*) \xrightarrow{p} H^n(k_s^*) \to H^{n+1}(\mathbf{F}_p) \to H^{n+1}(k_s^*) \xrightarrow{p} H^{n+1}(k_s^*) \to H^{n+2}(\mathbf{F}_p).$$

Suppose $\mathrm{cd}_p(G_k) \leqq n$. Then $H^{n+1}(\mathbf{F}_p) = H^{n+2}(\mathbf{F}_p) = 0$ by Proposition 3.7. Conditions (i) and (ii) are then clear, taking Proposition 3.5 into account. The converse can be proved in a similar way, from the fact that if G_p is a p-Sylow subgroup of G_k, then

$$H^r(G_p, k_s^*) = \mathrm{inv}\ \lim H^r(G_E, k_s^*),$$

the limit being taken over all finite separable extensions E of k of degree prime to p. The Galois groups G_E constitute precisely the set of open subgroups of G_k containing G_p, or its conjugates, which amounts to the same thing.

Corollary 6.5. *If $H^2(G_E, k_s^*) = 0$ for all finite separable extensions E of k, then $\mathrm{cd}(G_k) \leqq 1$.*

Proof. We have $H^1(G_e, k_s^*) = 0$ automatically, and we apply the theorem for the p-component when $p \neq$ char k. If $p =$ char k, then we saw in Theorem 6.1 that the cohomological dimension is $\leqq 1$.

The above corollary provides the announced criterion in terms of the Brauer group, because that is what $H^2(k_s^*)$ amounts to.

Theorem 6.6. *Let K be an extension of k. Then*

$$\mathrm{cd}_p(G_K) \leqq \mathrm{tr}\ \deg K/k + \mathrm{cd}_p(G_k).$$

(By definition, tr deg *is the transcendence degree.)*

Proof. If in a tower $K \supset K_1 \supset k$ the assertion is true for K/K_1 and for K_1/k, then it is true for K/k. We are therefore reduced to the cases when either K/k is algebraic, in which case G_K is a closed subgroup of G_k, and the assertion is trivial; or when K is a pure transcendental extension $K = k(x)$, in which case we have a field diagram as follows.

$$k(x) \xrightarrow{\ G_k\ } k^a(x) \xrightarrow{\ G_{k^a(x)}\ } k^a(x)_s$$

$$\uparrow \qquad\qquad \uparrow$$

$$k \xrightarrow[\ G_k\ ]{} k^a$$

By Tsen's theorem and the corollary of Theorem 6.3, we know that $\mathrm{cd}(G_{k^a(x)}) \leqq 1$. The tower theorem shows that

$$\mathrm{cd}(G_{k(x)}) \leqq \mathrm{cd}(G_k) + 1,$$

thus proving the theorem.

Theorem 6.7. *In the preceding theorem, there is equality if* $\mathrm{cd}_p(G_k) < \infty$ *($p \neq$ char k) and K is finitely generated over k.*

Proof. The assertion is again transitive in towers, and we are reduced either to the case of a finite algebraic extension, when we can apply Proposition 3.12, or to $K = k(x)$ purely transcendental. For this last case, we need a lemma.

Lemma 6.8. *Let G be of Galois type, T a closed normal subgroup such that* $\mathrm{cd}_p(T) \leqq 1$. *If* $\mathrm{cd}_p(G/T) \leqq n$, *then there is an isomorphism*

$$H^{n+1}(G, A) \approx H^n(G/T, H^1(T, A))$$

for all $A \in \mathrm{Galm}_{\mathrm{tor}}(G)$.

Proof. We have $H^r(T, A) = 0$ for $r > 1$ and the spectral sequence becomes an exact sequence

$$0 \to H^{n+1}(G/T, A^T) \to H^{n+1}(G, A) \to H^n(G/T, H^1(T, A))$$
$$\to H^{n+2}(G/T, A^T) \to 0,$$

whence the lemma follows.

Coming back to the theorem, put

$$G = G_{k(x)} \text{ and } T = G_{k(x)_s/k_s(x)}.$$

We refer to the diagram for Theorem 6.6. We may replace k be its extension corresponding to a Sylow subgroup of G_k, that is we may suppose that G_k is a p-group. We have $G_k = G/T$. Let us now take $A = \mathbf{F}_p$ in the lemma. Suppose

$$n = \mathrm{cd}_p(G/T) < \infty.$$

We must show that $H^{n+1}(G, A) \neq 0$. By the lemma, this amounts to showing that $H^n(G/T, H^1(T, \mathbf{F}_p)) \neq 0$. Since the p-th roots of unity are in k ($p \neq$ char k), Kummer theory shows that

$$H^1(T, F_p) = \mathrm{cont} \ \mathrm{hom}(T, \mathbf{F}_p)$$

is G/T-isomorphic to $k_s(x)^*/k_s(x)^{*p}$. The unique factorization in $k_s(x)$ shows that this group contains a subgroup G-isomorphic to $\mathbf{F}_p$. On the other hand, this group is a direct sum of its orbits under G/T, and one of these orbits is $\mathbf{F}_p$. Hence $H^{n+1}(G, \mathbf{F}_p) \neq 0$ as was to be shown.

The theorem we have just proved, and which occurs here at the end of the theory, historically arose at its beginning. Its conjecture and the sketch of its proof are due to Grothendieck.

CHAPTER VIII
Group Extensions

§1. Morphisms of extensions

Let G be a group and A an abelian group, both written multiplicatively. An **extension** of A by G is an exact sequence of groups

$$0 \to A \xrightarrow{i} U \xrightarrow{j} G \to 0.$$

We can then define an action of G on A. If we identify A as a subgroup of U, then U acts on A by conjugation. Since A is assumed commutative, it follows that elements of A act trivially, so $U/A = G$ acts on A.

For each $\sigma \in G$ and $a \in A$ we select an element $u_\sigma \in U$ such that $j u_\sigma = \sigma$, and we put

$$^\sigma a = [u_\sigma]a = u_\sigma a u_\sigma^{-1}.$$

Each element of U can be written uniquely in the form

$$u = a u_\sigma \text{ with } \sigma \in G \text{ and } a \in A.$$

Then there exist elements $a_{\sigma,\tau} \in A$ such that

$$u_\sigma u_\tau = a_{\sigma,\tau} u_{\sigma\tau},$$

and $(a_{\sigma,\tau})$ is a 2-cocycle of G in A. A different choice of u_σ would give rise to another cocycle, differing from the first one by a coboundary. Hence the cohomology class α of these cocycles is a well defined element of $H^2(G, A)$, determined by the extension, i.e. by the exact sequence.

Conversely, suppose given an element $\alpha \in H^2(G, A)$ with G given, and A abelian in $\mathrm{Mod}(G)$. Let $(a_{\sigma,\tau})$ be a cocycle representing α. We can then define an extension of A by G as follows. We let U be the set of pairs (a, σ) with $a \in A$ and $\sigma \in G$. We define multiplication in U by

$$(a, \sigma)(b, \tau) = (a\sigma b a_{\sigma,\tau}, \sigma\tau).$$

One verifies that U is a group, whose unit element is $(a_{e,e}^{-1}, e)$. The existence of the inverse of (a, σ) is determined at once from the definition of multiplication. Defining $j(a, \sigma) = \sigma$ gives a homomorphism of U onto G, whose kernel is isomorphic to A, under the correspondence

$$a \mapsto (aa_{e,e}^{-1}, e).$$

Thus we get a group extension of A by G.

Extensions of groups form a category, the morphisms being triplets of homomorphisms (f, F, φ) which make the following diagram commutative:

$$
\begin{array}{ccccccccc}
0 & \longrightarrow & A & \longrightarrow & U & \longrightarrow & G & \longrightarrow & 0 \\
& & {\scriptstyle f}\downarrow & & {\scriptstyle F}\downarrow & & {\scriptstyle \varphi}\downarrow & & \\
0 & \longrightarrow & B & \longrightarrow & V & \longrightarrow & H & \longrightarrow & 0
\end{array}
$$

We have the general notion of isomorphism in this category, but we look at the restricted notion of extensions U, U' of A by G (so the same A and G). Two such extensions will be said to be **isomorphic** if there exists an isomorphism $F : U \to U'$ making the following diagram commutative:

$$
\begin{array}{ccccc}
A & \longrightarrow & U & \longrightarrow & G \\
{\scriptstyle \mathrm{id}}\downarrow & & {\scriptstyle F}\downarrow & & {\scriptstyle \mathrm{id}}\downarrow \\
A & \longrightarrow & U' & \longrightarrow & G
\end{array}
$$

Isomorphism classes of extensions thus form a category, the morphisms being given by isomorphisms F as above.

Let (G, A) be a pair consisting of a group G and a G-module A. We denote by $E(G, A)$ the isomorphism classes of extensions of A by G. For G fixed, $A \mapsto E(G, A)$ is a functor on $\text{Mod}(G)$. We may summarize the discussion

Theorem 1.1. *On the category* $\text{Mod}(G)$, *the functors* $H^2(G, A)$ *and* $E(G, A)$ *are isomorphic, by the bijection established at the beginning of the section.*

Next, we state a general result providing the existence of the homomorphism F when pairs of homomorphisms (φ, f) are given, from a pair (G, A) to a pair (G', A').

Theorem 1.2. *Let* $G \approx U/A$ *and* $G' \approx U'/A'$ *be two extensions. Suppose given two homomorphisms*

$$\varphi : G \to G' \quad and \quad f : A \to A'.$$

There exists a homomorphism $F : U \to U'$ *making the diagram commutative:*

$$
\begin{array}{ccccc}
A & \xrightarrow{\ i\ } & U & \xrightarrow{\ j\ } & G \\
{\scriptstyle f}\downarrow & & {\scriptstyle F}\downarrow & & {\scriptstyle \varphi}\downarrow \\
A' & \xrightarrow[\ i'\ ]{} & U' & \xrightarrow[\ j'\ ]{} & G'
\end{array}
$$

if and only if:

(1) f *is a* G-homomorphism, *with* G *acting on* A' *via* φ.

(2) $f_*\alpha = \varphi^*\alpha$, *where* α, α' *are the cohomology classes of the two extensons respectively, and* f_*, φ^* *are the morphisms induced by the morphisms of pairs*

$$(\text{id}, f) : (G, A) \to (G, A') \quad and \quad (\varphi, \text{id}) : (G', A') \to (G, A').$$

Proof. We begin by showing that the conditions are necessary. Without loss of generality, we let i be an inclusion. Let (u_σ) and $(u'_{\sigma'})$ be representatives of σ and σ' respectively in G and G'. For $u = au_\sigma$ in U we find

$$F(u) = F(au_\sigma) = F(a)F(u_\sigma) = f(a)F(u_\sigma).$$

One sees that F is uniquely determined by the data $F(u_\sigma)$. We have

$$jF(u_\sigma) = \varphi j u_\sigma = \varphi\sigma = j' u'_{\varphi\sigma}.$$

Hence there exist elements $c_\sigma \in A'$ such that

$$F(u_\sigma) = c_\sigma u'_{\varphi\sigma}.$$

It follows that F is uniquely determined by the data (c_σ), which is a cochain of G in A'. Since F is a homomorphism, we must have

$$F(u_\sigma a u_\sigma^{-1}) = F(u_\sigma)f(a)f(u_\sigma)^{-1}$$
$$F(u_\sigma u_\tau) = F(u_\sigma)F(u_\tau).$$

These conditions imply:

(I) $f(\sigma a) = \varphi\sigma(fa)$ for $a \in A$ and $\sigma \in G$.

(II) $f a_{\sigma,\tau} = b_{\varphi\sigma,\varphi\tau}((\varphi\sigma)c_\tau c_{\sigma\tau}^{-1} c_\sigma)$,

for the cocycles $(a_{\sigma,\tau})$ and $(b_{\sigma',\tau'})$ associated to the representatives (u_σ) and (u'_σ). By definition, these two conditions express precisely the conditions (1) and (2) of the theorem. Conversely, one verifies that these conditions are sufficient by defining

$$F(a u_\sigma) = f(a)c_\sigma u'_{\varphi\sigma}.$$

This concludes the proof.

We also want to describe more precisely the possible F in an isomorphism class of extensions of A by G. We work more generally with the situation of Theorem 1.2. Let f, φ be fixed and let

$$F_1, F_2 : U \to U'$$

be homomorphisms which make the diagram of Theorem 1.2 commutative. We say that F_1 is **equivalent** to F_2 if they differ by an inner automorphism of U' coming from an element of A', that is there exists $a' \in A'$ such that

$$F_1(u) = a' F_2(u)a'^{-1} \quad \text{for all} \quad u \in U.$$

This equivalence is the weakest one can hope for.

Theorem 1.3. *Let f, φ be given as in Theorem 1.2. Then the equivalence classes of homomorphisms F as in this theorem form a principal homogeneous space of $H^1(G, A')$. The action of $H^1(G, A')$ on this space is defined as follows. Let (u_σ) be representatives of G in U, and (z_σ) a 1-cocycle of G in A'. Then*

$$(zF)(au_\sigma) = f(a)z_\sigma F(u_\sigma).$$

Proof. The straightforward proof is left to the reader.

Corollary 1.4. *If $H^1(G, A') = 0$, then two homomorphisms $F_1, F_2 : U \to U'$ which make the diagram of Theorem 1.2 commutative are equivalent.*

§2. Commutators and transfer in an extension.

Let G be a finite group and $A \in \mathrm{Mod}(G)$. We shall write A multiplicatively, and so we replace the trace by the norm $\mathbf{N} = \mathbf{N}_G$. We consider an extension of A by G,

$$0 \to A \xrightarrow{i} E \xrightarrow{j} G \to 0,$$

and we suppose without loss of generality that i is an inclusion. We fix a family of representatives (u_σ) of G in E, giving rise to the cocycle $(a_{\sigma,\tau})$ as in the preceding section. Its class is denoted by α. We let E^c be the commutator subgroup of E. The notations will remain fixed throughout this section.

Proposition 2.1. *The image of the transfer*

$$\mathrm{Tr} : E/E^c \to A$$

is contained in A^G, and one has:
(1) $\mathrm{Tr}(aE^c) = \prod_{\sigma \in G} u_\sigma a u_\sigma^{-1} = \mathbf{N}_G(a)$ *for $a \in A$.*
(2) $\mathrm{Tr}(u_\tau E^c) = \prod_{\sigma \in G} u_\sigma u_\tau u_{\sigma\tau}^{-1} = \prod_{\sigma \in G} a_{\sigma,\tau}$ *(the Nakayama map).*

Proof. These formulas are immediate consequences of the definition of the transfer.

Proposition 2.2. *One has $I_G A \subset E^c \cap A \subset A_N$. For the cup product relative to the pairing $\mathbf{Z} \times A \to A$, we have*

$$\alpha \cup \mathbf{H}^{-3}(G, \mathbf{Z}) = \aleph_G((E^c \cap A)/I_G A).$$

Proof. We have at once $\sigma a/a = u_\sigma a u_\sigma^{-1} a^{-1} \in E^c \cap A$. The other stated inclusion can be seen from the fact that tr is trivial on E^c, and applying Proposition 2.1. Now for the statement about the cup product, recall that a subgroup of an abelian group is determined by the group of characters $f : A \to \mathbf{Q}/\mathbf{Z}$ vanishing on the subgroup. A character $f : A \to \mathbf{Q}/\mathbf{Z}$ vanishes on $E^c \cap A$ if and only if we can extend f to a character of E/E^c, because

$$A/(E^c \cap A) \subset E/E^c.$$

The extension of a character can be formulated in terms of a commutative diagram such as those we considered previously, and of the existence of a map F, namely:

$$
\begin{array}{ccccc}
A & \longrightarrow & E & \longrightarrow & G \\
\downarrow{\scriptstyle f} & & \downarrow{\scriptstyle F} & & \downarrow \\
\mathbf{Q}/\mathbf{Z} & \longrightarrow & \mathbf{Q}/\mathbf{Z} & \longrightarrow & 0
\end{array}
$$

The existence of F is equivalent to the conditions:

(a) f is a G-homomorphism.
(b) $f_* \alpha = 0$.

From the definition of the cup product, we have a commutative diagram:

$$
\begin{array}{ccc}
\mathbf{H}^{-3}(\mathbf{Z}) \times \mathbf{H}^2(\mathbf{Q}/\mathbf{Z}) & \longrightarrow & \mathbf{H}^{-1}(\mathbf{Q}/\mathbf{Z}) = (\mathbf{Q}/\mathbf{Z})_N \\
\uparrow{\scriptstyle \mathrm{id}} \quad \uparrow{\scriptstyle f_*} & & \uparrow{\scriptstyle f_*} \\
\mathbf{H}^{-3}(\mathbf{Z}) \times \mathbf{H}^2(A) & \longrightarrow & \mathbf{H}^{-1}(A) = A_N/I_G A
\end{array}
$$

The duality theorem asserts that $\mathbf{H}^{-3}(\mathbf{Z})$ is dual to $\mathbf{H}^2(\mathbf{Q}/\mathbf{Z})$. In addition, the effect of f_* on $\mathbf{H}^{-1}(A)$ is induced by f on $A_N/I_G A$.

Suppose that f is a character of A vanishing on A_N. Then

$$f_*(\alpha \cup \mathbf{H}^{-3}(A)) = 0, \text{ and so } f_*(\alpha) \cup \mathbf{H}^{-3}(\mathbf{Z}) = 0.$$

Since $\mathbf{H}^{-3}(Z)$ is the character group of $\mathbf{H}^2(\mathbf{Q}/\mathbf{Z})$, we conclude that $f_*(\alpha) = 0$. The converse is proved in a similar way. This concludes the proof of Proposition 2.2.

In addition, Proposition 1.1 also gives:

Theorem 2.3. *Let*

$$0 \to A \xrightarrow{i} E \xrightarrow{j} G \to 0$$

be an extension, and $\alpha \in H^2(G, A)$ *its cohomology class. Then the following diagram is commutative:*

$$
\begin{array}{ccccccccc}
0 & \longrightarrow & A/E^c \cap A & \xrightarrow{\bar{i}} & E/E^c & \xrightarrow{\bar{j}} & G/G^c = \mathbf{H}^{-2}(G, \mathbf{Z}) & \longrightarrow & 0 \\
& & \downarrow \bar{\mathrm{N}} & & \downarrow \mathrm{Tr} & & \downarrow \alpha_{-2} & & \\
0 & \longrightarrow & \mathrm{N}A & \longrightarrow & A^G & \longrightarrow & \mathbf{H}^0(G, A) & \longrightarrow & 0
\end{array}
$$

where $\bar{\mathrm{N}}, \bar{i}, \bar{j}$ *are the homomorphisms induced by the norm, the inclusion, and* j *respectively; and* α_{-2} *denotes the cup product with* α *on* $H^{-2}(G, \mathbf{Z})$.

Proof. The left square is commutative because of the formula for the norm in Proposition 2.1(1). The transfer maps E/E^c into A^G by Proposition 2.1. The right square is commutative because the Nakayama map is an explicit determination of the cup product, and we can apply Proposition 2.1(2).

The next two corollaries are especially important in the application to class modules and class formations as in Chapter IX. They give conditions under which the transfer is an isomorphism.

Corollary 2.4. *Let* $E/A = G$ *be an extension with corresponding cohomology class* $\alpha \in H^2(G, A)$. *With the three homomorphisms*

$$\mathrm{Tr} : E/E^c \to A^G$$
$$\alpha_{-3} : \mathbf{H}^{-3}(G, \mathbf{Z}) \to \mathbf{H}^{-1}(G, A)$$
$$\alpha_{-2} : \mathbf{H}^{-2}(G, \mathbf{Z}) \to \mathbf{H}^0(G, A)$$

we get an exact sequence

$$0 \to \mathbf{H}^{-1}(G, A)/\mathrm{Im}\ \alpha_{-3} \to \mathrm{Ker}\ \mathrm{Tr} \to \mathrm{Ker}\ \alpha_{-2} \to 0$$

and an isomorphism

$$0 \to A^G/\mathrm{Im\ Tr} \to \mathbf{H}^0(G,A)/\mathrm{Im\ }\alpha_{-2} \to 0.$$

Proof. Chasing around diagrams.

Corollary 2.5. *If α_{-2} and α_{-3} are isomorphisms, then the transfer on A^G/NA is an isomorphism in Theorem 2.3.*

The situation of Corollary 2.5 is realized for class modules or class formations in Chapter IX.

§3. The deflation

Let G be a group and $A \in \mathrm{Mod}(G)$, written multiplicatively. Let E_G be an extension of A by G. Let N be a normal subgroup of G and $E_N = j^{-1}(N)$, so we have two exact sequences:

$$0 \to A \to E_G \xrightarrow{j} G \to 0$$

$$0 \to A \to E_N \to N \to 0.$$

Then E_N is an extension of A by N, and if $\alpha \in H^2(G,A)$ is the cohomology class of E_G then $\mathrm{res}_N^G(\alpha)$ is the cohomology class of E_N.

One sees that E_N is normal in E_G, and in fact $E_G/E_N \approx G/N$. We obtain an exact sequence

$$0 \to E_N \to E_G \to G/N \to 0.$$

Since E_N is not necessaily commutative, we factor by E_N^c to get the exact sequence

$$0 \to E_N/E_N^c \to E_G/E_N^c \to G/N \to 0$$

giving an extension of E_N/E_N^c by G/N, called the **factor extension** corresponding to the normal subgroup N of G. The group lattice is as follows.

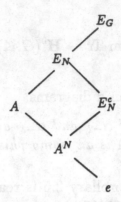

This factor extension corresponds to a cohomology class β in $H^2(G/N, E_N/E_N^c)$. We can take the transfer

$$\text{Tr} : E_N/E_N^c \to A^N,$$

which is a G/N-homomorphism, the operation of G/N on E_N/E_N^c being compatible with that of E_G/E_N^c. Consequently, there is a induced homomorphism

$$\text{Tr}_* : H^r(G/N, E^N/E_N^c) \to H^r(G/N, A^N).$$

The image of $\text{Tr}_*(\beta)$ depends only on α. Hence we get a map

$$\text{def} : H^2(G, A) \to H^2(G/N, A^N) \text{ such that } \alpha \mapsto \text{Tr}_*(\beta).$$

We call this map the **deflation**. It may not be a homomorphism, but we shall see that for G finite, it is. First:

Theorem 3.1. *Let S be a subgroup of a finite group G. Fix right coset representatives of S in G, and for $\sigma \in G$ let $\bar{\sigma}$ be the representative of $S\sigma$. Let $A \in \text{Mod}(G)$ and let $(a_{\sigma,\tau})$ be a 2-cocycle of G in A. Let E_G be the extension of A by S obtained from the restriction of this cocycle to S. Let (u_σ) be representatives of G in E_G. Let $\gamma(\sigma, \tau) = \bar{\sigma}\bar{\tau}\overline{\sigma\tau}^{-1}$. Then*

$$\text{Tr}_A^{E_S}(u_{\bar{\sigma}} u_{\bar{\tau}} u_{\overline{\sigma\tau}}^{-1}) = \prod_{\rho \in S} a_{\rho,\gamma}.$$

Proof. This comes directly from the formulas of Theorem 2.1.

Corollary 3.2. *Let G be a finite group, N normal in G. Let $A \in \mathrm{Mod}(G)$. Then on $H^2(G, A)$, we have*

$$\mathrm{inf}_G^{G/N} \circ \mathrm{def}_{G/N}^G = (N : e).$$

Proof. One computes with the explicit formulas on cocycles. Note that the group A being written multiplicatively, the expression $(N : e)$ on the right is really the map $\alpha \mapsto \alpha^{(N:e)}$ for $\alpha \in H^2(G, A)$.

Theorem 3.3. *Let G be a finite group and N a normal subgroup. Then the deflation is a homomorphism. If $\alpha \in H^2(G, A)$ is represented by the cocycle $(a_{\sigma,\tau})$, then $\mathrm{def}(\alpha)$ is represented by the cocycle*

$$\prod_{\rho \in N} \rho(a_{\bar{\sigma}, \bar{\tau}} a_{\gamma, \overline{\sigma\tau}}^{-1}) \prod_{\rho \in N} a_{\rho, \gamma} = \prod_{\rho \in N} a_{\bar{\sigma}\rho, \bar{\tau}} a_{\rho, \bar{\sigma}} a_{\rho, \overline{\sigma\tau}}^{-1}.$$

As in Theorem 3.1, $\bar{\sigma}$ denotes a fixed coset representative of the coset N^{σ}, and $\gamma = \bar{\sigma}\bar{\tau}\overline{\sigma\tau}^{-1} = \gamma(\sigma, \tau)$.

Proof. The proof is done by an explicit computation, using the explicit formula for the transfer in Theorem 3.1. The fact that the deflation is a homomorphism is then apparent from the expression on the right side of the equality. One also sees from this right expression that the expression on the left is well defined. The details are left to the reader.

CHAPTER IX
Class Formations

§1. Definitions

Let G be a group of Galois type, with a fundamental system of open neighborhoods of e consisting of open subgroups of finite index $U, V, \ldots$. Let $A \in \operatorname{Galm}(G)$ be a Galois module. We then say that the pair (G, A) is a **class formation** if it satisfies the following two axioms:

CF 1. For each open subgroup V of G one has $H^1(V, A) = 0$.

Because of the inflation-restriction exact sequence in dimension 1, this axiom is equivalent to the condition that for all open subgroups U, V with U normal in V, we have

$$H^1(V/U, A^U) = 0.$$

Example. If k is a field and K is a Galois extension of k with Galois group G, then (G, K^*) satisfies the axiom **CF 1**.

By **CF 1**, it follows that the inflation-restriction sequence is exact in dimension 2, and hence that the inflations

$$\inf : H^2(V/U, A^U) \to H^2(V, A)$$

are monomorphisms for V open, U open and normal in V. We may therefore consider $H^2(V, A)$ as the union of the subgroups

$H^2(V/U, A^U)$. It is by definition the **Brauer group** in the preceding example. The second axiom reads:

CF 2. For each open subgroup V of G we are given an embedding

$$\mathrm{inv}_V : H^2(V, A) \to \mathbf{Q}/\mathbf{Z} \text{ denoted } \alpha \mapsto \mathrm{inv}_V(\alpha),$$

called the **invariant**, satisfying two conditions:

(a) If $U \subset V$ are open and U is normal in V, of index n in V, then inv_V maps $H^2(V/U, A^U)$ onto the subgroup $(\mathbf{Q}/\mathbf{Z})_n$ consisting of the elements of order n in $\mathbf{Q}/\mathbf{Z}$.

(b) If $U \subset V$ are open subgroups with U of index n in V, then

$$\mathrm{inv}_V \circ \mathrm{res}_U^V = n.\mathrm{inv}_V.$$

We note that if $(G : e)$ is divisible by every positive integer m, then inv_G maps $H^2(G, A)$ onto $\mathbf{Q}/\mathbf{Z}$, i.e.

$$\mathrm{inv}_G : H^2(G, A) \to \mathbf{Q}/\mathbf{Z}$$

is an isomorphism. This is the case in both local and global class field theory over number fields:

In the local case, A is the multiplicative group of the algebraic closure of a p-adic field, and G is the Galois group.

In the global case, A is the direct limit of the groups of idele classes. On the other hand, if G is finite, then of course inv_G maps $H^2(G, A)$ only on $(\mathbf{Q}/\mathbf{Z})_n$, with $n = (G : e)$.

Let G be finite, and (G, A) a class formation. Then A is a class module. But for a class formation, we are given an additional structure, namely the specific fundamental elements $\alpha \in H^2(G, A)$ whose invariant is $1/n \pmod{\mathbf{Z}}$.

Let (G, A) be a class formation and $U \subset V$ open subgroups with U normal in V. The element $\alpha \in H^2(V/U, A^U) \subset H^2(V, A)$, whose V-invariant $\mathrm{inv}_V(\alpha)$ is $1/(V : U)$, will be called **the fundamental class** of $H^2(V/U, A^U)$, or by abuse of language, of V/U.

Proposition 1.1. *Let $U \subset V \subset W$ be three open subgroups of G, with U normal in W. If α is the fundamental class of W/U then $\mathrm{res}_V^W(\alpha)$ is the fundamental class of V/U.*

Proof. This is immediate from **CF 2(b)**.

Corollary 1.2. *Let (G, A) be a class formation.*

(i) *Let V be an open subgroup of G. Then (V, A) is a class formation, and the restriction*

$$\mathrm{res} : H^2(G, A) \to H^2(V, A)$$

is surjective.

(ii) *Let N be a closed normal subgroup. Then $(G/N, A^N)$ is a class formation, if we define the invariant of an element in $H^2(VN/N, A^N)$ to be the invariant of its inflation in $H^2(VN, A)$.*

Proof. Immediate.

Proposition 1.3. *Let (G, A) be a class formation and let V be an open subgroup of G. Then:*

(i) *The transfer preserves invariants, that is for $\alpha \in H^2(V, A)$ we have*

$$\mathrm{inv}_G \; \mathrm{tr}_G^V(\alpha) = \mathrm{inv}_V(\alpha).$$

(ii) *Conjugation preserves invariants, that is for $\alpha \in H^2(V, A)$ we have*

$$\mathrm{inv}_{V[\sigma]}(\sigma_* \alpha) = \mathrm{inv}_V(\alpha)$$

Proof. Since the restriction is surjective, the first assertion follows at once from **CF 2** and the formula

$$\mathrm{tr} \circ \mathrm{res} = (G : V).$$

As for the second, we recall that σ_* is the identity on $H^2(G, A)$. Hence

$$\mathrm{inv}_{V[\sigma]} \circ \sigma_* \circ \mathrm{res}_V^G = \mathrm{inv}_{V[\sigma]} \mathrm{res}_{V[\sigma]}^G \circ \sigma_*$$

$$= (G : V[\sigma]) \mathrm{inv}_G \circ \sigma_*$$

$$= (G : V) \mathrm{inv}_G$$

$$= \mathrm{inv}_V \mathrm{res}_V^G.$$

Since the restriction is surjective, the proposition follows.

Theorem 1.4. *Let G be a finite group and (G, A) a class formation. Let α be the fundamental element of $H^2(G, A)$. Then the cup product*

$$\alpha_r : \mathbf{H}^r(G, \mathbf{Z}) \to \mathbf{H}^{r+2}(G, A)$$

is an isomorphism for all $r \in \mathbf{Z}$.

Proof. For each subgroup G' of G let α' be the retriction to G', and let α'_r the cup product taken on the G'-cohomology. By the triplets theorem, it will suffice to prove that α'_r satisfies the hypotheses of this theorem in three successive dimensions, which we choose to be dimensions $-1, 0$, and $+1$.

For $r = -1$, we have $H^1(G, A) = 0$ so α'_{-1} is surjective.

For $r = 0$, we note that $\mathbf{H}^0(G', \mathbf{Z})$ has order $(G' : e)$ which is the same order as $H^2(G', A)$. We have trivially

$$\alpha'_0(\varkappa(1)) = \alpha',$$

which shows that α'_0 is an isomorphism.

For $r = 1$, we simply note that $H^1(G, \mathbf{Z}) = 0$ since G is finite and the action on $\mathbf{Z}$ is trivial. This concludes the proof of the theorem.

Next we make explicit some commutativity relations for restriction, transfer, inflation and conjugation relative to the natural isomorphism of $\mathbf{H}^r(G, A)$ with $\mathbf{H}^{r-2}(G, \mathbf{Z})$, cupping with α.

Proposition 1.5. *Let G be a finite group and (G, A) a class formation. let $\alpha \in H^2(G, A)$ be a fundamental element and α' its restriction to G' for a subgroup G' of G. Then for each pair of vertical arrows pointing in the same direction, the following diagram is commutative.*

$$
\begin{array}{ccc}
\mathbf{H}^r(G, \mathbf{Z}) & \xrightarrow{\;\alpha_r\;} & \mathbf{H}^{r+2}(G, A) \\
\text{res} \big\updownarrow \text{tr} & & \text{res} \big\updownarrow \text{tr} \\
\mathbf{H}^r(G', \mathbf{Z}) & \xrightarrow[\;\alpha'_r\;]{} & \mathbf{H}^{r+2}(G', A)
\end{array}
$$

Proof. This is just a special case of the general commutativity relations.

Proposition 1.6. *Let G be a finite group and (G, A) a class formation. Let U be normal in G. Let $\alpha \in H^2(G, A)$ be the fundamental element, and $\bar{\alpha}$ the fundamental element for G/U. Then the following diagram is commutative for $r \geqq 0$.*

$$
\begin{array}{ccc}
H^r(G/U, \mathbf{Z}) & \xrightarrow{\ \bar{\alpha}_r\ } & H^{r+2}(G/U, A^U) \\
{\scriptstyle (U:e)\text{inf}} \big\uparrow & & \big\uparrow {\scriptstyle \text{inf}} \\
H^r(G, \mathbf{Z}) & \xrightarrow[\ \alpha_r\ ]{} & H^{r+2}(G, A)
\end{array}
$$

Proof. This is just a special case of the rule

$$\inf(\alpha \cup \beta) = \inf(\alpha) \cup \inf(\beta).$$

We have to observe that we deal with the ordinary functor H in dimension $r \geq 0$, differing from the special one only in dimension 0, because the inflation is defined only in this case. The left homomorphism is $(U : e)$inf for the inflation, given by the inclusion. Indeed, we have

$$(G : e) = (G : U)(U : e),$$

so $\inf(\bar{\alpha}) = (U : e)\alpha$, and we can apply the above rule.

Finally, we consider some isomorphisms of class formations. Let (G, A) and (G', A') be class formations. An **isomorphism**

$$(\lambda, f) : (G', A') \to (G, A)$$

consists of a pair isomorphism $\lambda : G \to G'$ and $f : A' \to A$ such that

$$\operatorname{inv}_G(\lambda, f)_*(\alpha') = \operatorname{inv}_{G'}(\alpha') \text{ for } \alpha' \in H^2(G', A').$$

From such an isomorphism, we obtain a commutative diagram for U normal in V (subgroups of G):

$$
\begin{array}{ccc}
H^r(V/U, \mathbf{Z}) & \xrightarrow{\ \alpha_r\ } & H^{r+2}(V/U, A^U) \\
{\scriptstyle (\lambda,1)_*} \big\downarrow & & \big\downarrow {\scriptstyle (\lambda,f)_*} \\
H^r(\lambda V/\lambda U, \mathbf{Z}) & \xrightarrow[\ \alpha'_r\ ]{} & H^{r+2}(\lambda V/\lambda U, A'^{\lambda U})
\end{array}
$$

where α, α' denote the fundamental elements in their respective H^2.

Conjugation is a special case, made explicit in the next proposition.

Proposition 1.7. *Let (G, A) be a class formation, and $U \subset V$ two open subgroups with U normal in V. Let $\tau \in G$ and α the fundamental element in $H^2(V/U, A^U)$. Then the following diagram is commutative.*

$$
\begin{array}{ccc}
H^r(V/U, \mathbf{Z}) & \xrightarrow{\ \alpha_r\ } & H^{r+2}(V/U, A^U) \\
{\scriptstyle \tau_*}\downarrow & & \downarrow{\scriptstyle \tau_*} \\
H^r(V[\tau]/U[\tau], \mathbf{Z}) & \xrightarrow[\tau_* \alpha_r]{} & H^{r+2}(V[\tau]/U[\tau], A^{U[\tau]}).
\end{array}
$$

§2. The reciprocity homomorphism

We return to Theorem 8.7 of Chapter IV, but with the additional structure of the class formation. From that theorem, we know that if (G, A) is a class formation and G is finite, then G/G^c is isomorphic to $A^G/S_G A = \mathbf{H}^0(G, A)$. The isomorphism can be realized in two ways. First, directly, and second by duality. Here we start with the duality. We have a bilinear map

$$
A^G \times \hat{G} \to H^2(G, A)
$$

given by

$$
(a, \chi) \mapsto \varkappa(a) \cup \delta\chi.
$$

Following this with the invariant, we obtain a bilinear map

$$
(a, \chi) \mapsto \mathrm{inv}_G(\varkappa(a) \cup \delta\chi) \text{ of } A^G \times \hat{G} \to \mathbf{Q}/\mathbf{Z},
$$

whose kernel on the left is $S_G A$ and whose kernel on the right is trivial. Hence $A^G/S_G A \approx G/G^c$, both groups being dual to $\hat{G}$. We recall the commutative diagram:

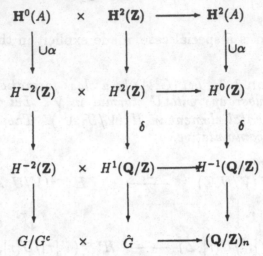

which we apply to the fundamental cocycle $\alpha \in H^2(G, A)$, with $n = (G : e)$ and $\mathrm{inv}_G(\alpha) = 1/n$. We have

$$\varkappa(1) \cup \alpha = \alpha \text{ and } \mathrm{inv}_G(\varkappa(1) \cup \alpha) = \mathrm{inv}_G(\alpha) = 1/n.$$

Thus using the invariant from a class formation, at a finite level, we obtain the following fundamental result.

Theorem 2.1. *Let G be a finite group and (G, A) a class formation. For $a \in A^G$ let σ_a be the element of G/G^c corresponding to a under the above isomorphism. Then for all characters χ of G we have*

$$\chi(\sigma_a) = \mathrm{inv}_G(\varkappa(a) \cup \delta\chi).$$

An element $\sigma \in G$ is equal to σ_a if and only if for all characters χ,

$$\chi(\sigma) = \mathrm{inv}_G(\varkappa(a) \cup \delta\chi).$$

The map $a \mapsto \sigma_a$ induces an isomorphism $A^G/S_G A \approx G/G^c$.

The element σ_a in the theorem will also be denoted by (a, G).

Let now G be of Galois type and let (G, A) be a class formation. Then we have a bilinear map

$$H^0(G, A) \times H^1(G, \mathbf{Q}/\mathbf{Z}) \to H^2(G, A), \text{ i.e. } A^G \times \hat{G} \to H^2(G, A)$$

with the ordinary functor H^0, by the formula

$$(a, \chi) \mapsto a \cup \delta\chi,$$

where we identify a character χ with the corresponding element of $H^1(G, \mathbf{Q}/\mathbf{Z})$, and we identify $H^0(G, A)$ with A^G. Since inflation commutes with the cup product, we see that if U is normal open in G, then the following diagram is commutative:

$$
\begin{array}{ccccc}
H^0(G, A) & \times & H^1(G, \mathbf{Q}/\mathbf{Z}) & \longrightarrow & H^2(G, A) \\
\uparrow^{\inf} & & \uparrow^{\inf} & & \uparrow^{\inf} \\
H^0(G/U, A^U) & \times & H^1(G/U, \mathbf{Q}/\mathbf{Z}) & \longrightarrow & H^2(G/U, A^U)
\end{array}
$$

The inflation on the far left is simple the inclusion $A^U \subset A$, and the inflation in the middle is that of characters.

In particular, to each element $a \in A^G$ we obtain a character of $H^1(G, \mathbf{Q}/\mathbf{Z})$ given by

$$\chi \mapsto \operatorname{inv}_G(a \cup \delta\chi).$$

We consider $H^1(G, \mathbf{Q}/\mathbf{Z})$ as a discrete group. Its character group is G/G^c, according to Pontrjagin duality between discrete and compact groups, but G^c now denotes the closure of the commutator group. Thus we obtain a homomorphism

$$\operatorname{rec}_G : A^G \to G/G^c$$

which we call the **reciprocity homomorphism**, characterized by the property that for U open normal in G, and $a \in A^G$ we have

$$\operatorname{rec}_{G/U}(a) = (a, G/U) = (a, G/G^c U).$$

Similarly, we may replace U be any normal closed subgroup of G. This is called the **consistency** of the reciprocity mapping. As when G is finite, we denote

$$\operatorname{rec}_G(a) = (a, G).$$

The next theorem is merely a formal summary of what precedes for finite factor groups, and the consistency.

Theorem 2.2. *Let G be a group of Galois type and (G, A) a class formation. Then there exists a unique homomorphism*

$$\mathrm{rec}_G : A^G \to G/G^c \quad \text{denoted} \quad a \mapsto (a, G)$$

satisfying the property

$$inv_G(a \cup \delta\chi) = \chi(a, G)$$

for all characters χ of G.

Recall that if $\lambda : G_1 \to G_2$ is a group homomorphism, then λ induces a homomorphism

$$\lambda^c : G_1/G_1^c \to G_2/G_2^c.$$

This also holds for a continuous homomorphism of groups of Galois type, where G^c denotes the closure of the commutator group.

The next theorem summarizes the formalism of class formation theory and the reciprocity mapping.

Theorem 2.3. *Let G be a group of Galois type and (G, A) a class formation.*

(i) *If $a \in A^G$ and S is a closed normal subgroup with factor group $\lambda : G \to G/S$, then $\mathrm{rec}_{G/S} = \lambda^c \circ \mathrm{rec}_G$, that is*

$$(a, G/T) = \lambda^c(a, G).$$

(ii) *Let V be an open subgroup of G. Then $\mathrm{rec}_V = \mathrm{Tr}_V^G \circ \mathrm{rec}_G$, that is for $a \in A^G$,*

$$(a, V) = \mathrm{Tr}_V^G(a, G).$$

(iii) *Again let V be an open subgroup of G and let $\lambda : V \to G$ be the inclusion. Then $\mathrm{rec}_G \circ S_G^V = \lambda^c \circ \mathrm{rec}_V$, that is for $a \in A^V$,*

$$(S_G^V(a), G) = \lambda^c(a, V).$$

(iv) *Let V be an open subgroup of G and $a \in A^V$. Let $\tau \in G$. Then*

$$(\tau a, V^\tau) = (a, V)^\tau.$$

These properties are called respectively **consistency**, **transfer**, **translation**, and **conjugation** for the reciprocity mapping.

Proof. The consistency property is just the commutativity of inflation and cup product. We already used it when we defined the symbol (a, G) for G of Galois type. The other properties are proved by reducing them to the case when G is finite. For instance, let us consider (ii). To show that two elements of V/V^c are equal, it suffices to prove that for every character $\chi : V \to \mathbf{Q}/\mathbf{Z}$ the values of χ on these two elements are equal. To do this, there exists an open normal subgroup U of G with $U \subset V$ such that $\chi(U) = 0$. Let $\bar{G} = G/U$. Then the following diagram is commutative:

$$
\begin{array}{ccccc}
G/G^c & \xrightarrow{\ \mathrm{Tr}\ } & V/V^c & \xrightarrow{\ \chi\ } & \mathbf{Q}/\mathbf{Z} \\
\downarrow & & \downarrow & & \downarrow \\
\bar{G}/\bar{G}^c & \xrightarrow[\ \mathrm{Tr}\ ]{} & \bar{V}/\bar{V}^c & \xrightarrow[\ \chi\ ]{} & \mathbf{Q}/\mathbf{Z}
\end{array}
$$

the vertical maps being canonical. Furthermore, by consistency, we have

$$(a, G)U = (a, G/U) = (a, \bar{G}).$$

This reduces the property to the finite case $\bar{G}$.

But when G is finite, then we can also write

$$(a, G) = \sigma \Leftrightarrow \varkappa_G(a) = \zeta_\sigma \cup \alpha,$$

where ζ_σ is the element of $\mathbf{H}^{-2}(G, \mathbf{Z}) \approx G/G^c$ corresponding to σ, and α is the fundamental class. The restriction $\mathrm{res}_V^G(\alpha) = \alpha'$ is the fundamental class of $H^2(V, A)$, and we know that

$$\zeta_{\mathrm{Tr}(\sigma)} \cup \alpha' = \mathrm{res}_V^G(\zeta_\sigma \cup \alpha).$$

Since

$$\mathrm{res}_V^G(\varkappa_G(a)) = \varkappa_V(a),$$

one sees that $\mathrm{Tr}(a, G) = (a, V)$.

For (iii), note that the diagram is commutative,

$$
\begin{array}{ccccc}
V/V^c & \xrightarrow{\quad} & G/G^c & \xrightarrow{\ \chi\ } & \mathbf{Q}/\mathbf{Z} \\
\downarrow & & \downarrow & & \downarrow \\
\bar{V}/\bar{V}^c & \xrightarrow{\quad} & \bar{G}/\bar{G}^c & \xrightarrow[\ \chi\ ]{} & \mathbf{Q}\mathbf{Z}
\end{array}
$$

where as previously $U \subset V$ is normal in G and $\bar{G} = G/U, \bar{V} = V/U$. This reduces the property to the case when G is finite. In this case, let

$$\lambda^c : V/V^c \to G/G^c$$

be the homomorphism induced by inclusion. Let α be the fundamental class of (G, A). Then $\mathrm{res}_V^G(\alpha) = \alpha'$ is the fundamental class of (V, A). The transfer and cup product are related by the formula

$$\mathrm{tr}(\zeta_\tau \cup \alpha') = \zeta_{\lambda\tau} \cup \alpha.$$

But the transfer amounts to the trace on $H^0(V, A) = A^V$, so the assertion is proved.

The fourth property is just a transport of structure for algebraically defined notions and relations.

We state one more property somewhat different from the others.

Theorem 2.4. Limitation Theorem. *Let G be of Galois type, V an open subgroup, and (G, A) a class formation. Then the image of $\mathbf{S}_G^V(A^V)$ by the reciprocity mapping rec_G is contained in VG^c/G^c, and we have an isomorphism induced by rec_G, namely*

$$\mathrm{rec}_G : A^G/\mathbf{S}_G^V A^V \xrightarrow{\approx} G/VG^c.$$

Proof. The first assertion is Property (iii) of Theorem 2.3. Conversely, since V is open, we may assume without loss of generality that G is finite. In this case, there exists $b \in A^V$ such that $\lambda^c(b, V) = (a, G)$. By this same Property (iii), this is equal to $(\mathbf{S}_G^V(b), G)$. But we know that the kernel of rec_G is equal to $\mathbf{S}_G A$. Hence a and $\mathbf{S}_G^V(b)$ are congruent mod $\mathbf{S}_G(A)$. Since

$$\mathbf{S}_G(A) \subset \mathbf{S}_G^V(A),$$

we have proved the theorem.

Corollary 2.5. *Let G be finite and (G, A) a class formation. Let $G' = G/G^c$ and $A' = A^{G^c}$. Then $\mathbf{S}_G(A) = \mathbf{S}_{G'}(A')$ and $\mathrm{rec}_G, \mathrm{rec}_{G'}$ are equal, their kernels being $\mathbf{S}_G(A)$.*

Note that $G' = G/G^c$ can be written G^{ab}, and can be viewed as the maximal abelian quotient of G in Corollary 2.5. The corollary shows that the information in the reciprocity mapping is entirely concerned with this maximal abelian quotient.

Theorem 2.6. *Let G be abelian of Galois type. Let (G, A) be a class formation. Then the open subgroups V of G are in bijection with the subgroups of A of the form $\mathbf{S}_G^V(A^V)$, called the* **trace group**. *If we denote this subgroup by B_V, and U is an open subgroup of G, then $U \subset V$ if and only if $B_V \subset B_U$, and $B_{UV} = B_U \cap B_V$. If in addition B is a subgroup of A^G such that $B \supset B_V$ for some open subgroup V of G, then there exists U open subgroup of G such that $B = B_U$.*

Proof. All the assertions are special cases of what has previously been proved, except possibly for the last one. But for this one, one may suppose G finite and consider $(G/V, A^V)$ instead of (G, A). We let $U = \mathrm{rec}_G(B)$, and we find an isomorphism $B/\mathbf{S}_G(A) \approx U$, to which we apply Theorem 2.4 to conclude the proof.

A subgroup B of A^G will be called **admissible** if there exists V open subgroup of G such that $B = \mathbf{S}_B^V(A^G)$. We then write $B = B_V$. The next result is an immediate consequence of Theorem 2.6 and the basic properties of the reciprocity map.

Corollary 2.7. *Let G be a group of Galois type and (G, A) a class formation. Let $B \subset A^G$ be admissible, $B = B_U$, and suppose U normal, G/U abelian. Let V be an open subgroup of G and put*

$$C = (\mathbf{S}_G^V)^{-1}(B),$$

so C is a subgroup of A^V. Then C is admissible for the class formation (V, A), and C corresponds to the subgroup $U \cap V$ of V.

In the next section, we discuss in greater detail the relations between class formations and group extensions. However, we can already formulate the theorem of Shafarevich-Weil. Note that if G is of Galois type and U is open normal in G, then U/U^c is a Galois module for G, or in other words, U^c is normal in G. Consequently, G/U acts on U/U^c, and we obtain a group extension

$$(1) \qquad\qquad 0 \to U/U^c \to G/U^c \to G/U \to 0.$$

If in addition (G, A) is a class formation, then the reciprocity mapping

$$\mathrm{rec}_U : A^U \to U/U^c$$

is a G/U-homomorphism.

Theorem 2.8 (Shafarevich-Weil). *Let G be of Galois type, U open normal in G, and (G, A) a class formation. Then*

$$\mathrm{rec}_{U^*} : H^2(G/U, A^U) \to H^2(G/U, U/U^c)$$

maps the fundamental class on the class of the group extension (1). There exists a family of coset representatives $(\bar{\sigma})_{\sigma \in G}$ of U in G such that if $a_{\bar{\sigma}, \bar{\tau}}$ is a cocycle representing α, then

$$(a_{\bar{\sigma}, \bar{\tau}}, U) = \bar{\sigma}\bar{\tau}\overline{\sigma\tau}^{-1} U^c.$$

Proof. Let $V \subset U$ be open normal in G. Ultimately, we let V tend to e. By the deflation operation of Chapter VIII, Theorem 3.2, there exists a cocycle $b_{\bar{\sigma}, \bar{\tau}}$ representing the fundamental class of $H^2(G/V, A^V)$ and representatives $\bar{\sigma}$ of U/V such that

$$(2) \qquad a_{\bar{\sigma}, \bar{\tau}} = \mathbf{S}_{U/V}(b_{\bar{\sigma}, \bar{\tau}}/b_{\gamma(\sigma, \tau), \overline{\sigma\tau}}) \prod_{\rho \in U/V} b_{\rho, \gamma(\sigma, \tau)},$$

where $\gamma(\sigma, \tau) = \bar{\sigma}\bar{\tau}\overline{\sigma\tau}^{-1}$. Therefore, we find

$$(a_{\bar{\sigma}, \bar{\tau}}U/V) = \bar{\sigma}\bar{\tau}\overline{\sigma\tau}^{-1} V U^c.$$

We take a limit over V as follows. let $C_1, \ldots, C_m$ be the cosets of U in G. They are closed and compact. The product space $(C_1, \ldots, C_m)$ is compact. If $\bar{\sigma}_1, \ldots, \bar{\sigma}_m$ are representatives of U/V satisfying (2), then any representatives of the cosets $\bar{\sigma}_1 V, \ldots, \bar{\sigma}_m V$ will also satisfy (2). The subset $(\bar{\sigma}_1 V, \ldots \bar{\sigma}_m V)$ is closed in $(C_1, \ldots, C_m)$. From the consistency of the reciprocity map, these subsets have the finite intersection property. Hence their intersection taken over all V is not empty, and there exists representatives $\bar{\sigma}$ of the cosets of U in G which all give the same $a_{\bar{\sigma}, \bar{\tau}}$. The theorem is now clear.

§3. Weil groups

Let G be a group of Galois type and (G, A) a class formation. At the end of the preceding section, we saw the exact sequence

$$0 \to U/U^c \to G/U^c \to G/U \to 0$$

for every open subgroup U normal in G. Furthermore, U/U^c is isomorphic to the factor group $A^U/\mathbf{S}_G^U(A)$. We now seek an extension X of A^U by G/U and a commutative diagram

$$
\begin{array}{ccccccccc}
0 & \longrightarrow & A^u & \longrightarrow & X & \longrightarrow & G/U & \longrightarrow & 0 \\
 & & \big\downarrow{\scriptstyle rec_U} & & \big\downarrow & & \big\downarrow{\scriptstyle id} & & \\
0 & \longrightarrow & U/U^c & \longrightarrow & G/U^c & \longrightarrow & G/U & \longrightarrow & 0
\end{array}
$$

satisfying various properties made explicit below. The problem will be solved in the following discussion.

We start first with the finite case, so let G be finite. By a **Weil group** for (G, A) we mean a triple $(E, g, \{f_U\})$, consisting of a group E and a surjective homomorphism

$$E \xrightarrow{g} G \to 0$$

(so a group extension) such that, if we put $E_U = g^{-1}(U)$ for U an open subgroup of G (and so $E = E_G$), then f_U is an isomorphism

$$f_U : A^U \to E_U/E_U^c.$$

These data are assumed to satisfy four axioms:

W 1. For each pair of open subgroups $U \subset V$ of G, the following diagram is commutative:

$$
\begin{array}{ccc}
A^U & \xrightarrow{\ f_U\ } & E_U/E_U^c \\
{\scriptstyle inc}\big\uparrow & & \big\uparrow{\scriptstyle Tr} \\
A^V & \xrightarrow[\ f_V\ ]{} & E_V/E_V^c
\end{array}
$$

W 2. For $x \in E_G$ and every open subgroup U of G, the diagram is commutative:

$$
\begin{array}{ccc}
A^U & \xrightarrow{\ f_U\ } & E_U/E_U^c \\
\big\downarrow & & \big\downarrow \\
A^{U[x]} & \xrightarrow[\ f_{U[x]}\ ]{} & E_{U[x]}/E_{U[x]}^c
\end{array}
$$

The vertical isomorphisms are the natural ones arising from x.

Let $U \subset V$ be open subgroups of G, and U normal in V. Then we have a canonical isomorphism

$$E_V/E_U \approx V/U$$

and an exact sequence

(3) $$0 \to E_U/E_U^c \to E_V/E_U^c \to V/U \to 0.$$

Then V/U acts on E_U/E_U^c and **W 2** guarantees that f_U is a G/U-isomorphism. This being the case we can formulate the third axiom.

W 3. Let $f_{U_*} : H^2(V/U, A^U) \to H^2(V/U, E_U/E_U^c)$ be the induced homomorphism. Then the image of the fundamental class of $(V/U, A^U)$ is the class corresponding to the group extension defined by the exact sequence (3).

Finally we have a separation condition.

W 4. One has $E_e^c = e$, in other words the map $f_e : A \to E_e$ is an isomorphism.

Theorem 3.1. *Let G be a finite group and (G, A) a class formation. Then there exists a Weil group for (G, A). Its uniqueness will be described in the subsequent theorem.*

Proof. Let E_G be an extension of A by G,

$$0 \to A \to E_G \xrightarrow{g} G \to 0$$

corresponding to the fundamental class in $H^2(G, A)$. This extension is uniquely determined up to inner automorphisms by elements of A, because $H^1(G, A)$ is trivial (Corollary 1.4 of Chapter VIII), and we have an isomorphism

$$f_e : A \to E_e,$$

so **W 4** is satisfied.

For each $U \subset G$, we let $E_U = g^{-1}(U)$, the extreme cases being given by A and E_G. Thus we have an exact sequence

$$0 \to A \to E_U \to U \to 0$$

of subextension, and its class in $H^2(U, A)$ is the restriction of the fundamental class, i.e. it is a fundamental class for (U, A).

Consequently, if U is normal in G, we may form the factor extension

$$0 \to E_U/E_U^c \to E_G/E_U^c \to G/U \to 0.$$

By Corollary 2.5 of Chapter VIII, we know that the transfer

$$\mathrm{Tr} : E_U/E_U^c \to A^U$$

is an isomorphism, and one sees at once that it is a G/U-isomorphism. Its inverse gives us the desired map

$$f_U : A^U \to E_U/E_U^c.$$

It is now easy to verify that the objects $(E_G, g, \{f_U\})$ as defined above form a Weil group. The Axioms **W 1**, **W 2**, **W 4** are immediate, taking into account the transitivity of the transfer and its functoriality. For **W 3**, we have to consider the deflation. In light of the "functorial" definition of E_G, one may suppose that $V = G$ in axiom **W 3**. If α is the fundamental class in $H^2(G, A)$, then $(U : e)\alpha$ is the inflation of the fundamental class in $(G/U, A^U)$. By Corollary 3.2 of Chapter VIII, one sees that the deflation of the fundamental class of (G, A) to $(G/U, A^U)$ is the fundamental class of $(G/U, A^U)$. Since f_U is the inverse of the transfer, one sees from the definition of the deflation that axiom **W 3** is satisfied. This concludes the proof of existence.

We now consider the uniqueness of a Weil group. Suppose G finite, and let (G, A) be a class formation. let $(E, g, \{f_U'\})$ be two Weil groups. An **isomorphism** φ of the first on the second is a group isomorphism

$$\varphi : E \to E'$$

satisfying the following conditions:

ISOW 1. The diagram is commutative:

$$
\begin{array}{ccc}
E & \xrightarrow{\;g\;} & G \\
\varphi \downarrow & & \downarrow \mathrm{id} \\
E' & \longrightarrow & G.
\end{array}
$$

From **ISOW 1** we see that $\varphi(E_U) = E'_U$ for all open subgroups U, whence an isomorphism

$$\varphi_U^c : E_U/E_U^c \to E'_U/E'^c_U.$$

The second condition then reads:

ISOW 2. The diagram

$$
\begin{array}{ccc}
A^U & \xrightarrow{f_U} & E_U/E_U^c \\
{\scriptstyle \mathrm{id}}\downarrow & & \downarrow{\scriptstyle \varphi_U^c} \\
A^U & \xrightarrow[f'_U]{} & E'_U/E'^c_U
\end{array}
$$

is commutative for all open subgroups U of G.

Theorem 3.2. *Let G be a finite group. Two Weil groups associated to a class formation (G, A) are isomorphic. Such an isomorphism is uniquely determined up to an inner automorphism of E' by elements of E'_e.*

Proof. Let φ be an isomorphism. The following diagram is commutative by definition.

$$
\begin{array}{ccccccccc}
0 & \longrightarrow & A & \xrightarrow[\approx]{f_e} & E_e & \longrightarrow & E & \longrightarrow & G & \longrightarrow & 0 \\
& & {\scriptstyle \mathrm{id}}\downarrow & & \downarrow & & \downarrow{\scriptstyle \varphi} & & \downarrow & & \\
0 & \longrightarrow & A & \xrightarrow[f'_e]{\approx} & E'_e & \longrightarrow & E' & \longrightarrow & G & \longrightarrow & 0.
\end{array}
$$

Conversely, we claim that any homomorphism φ which makes this diagram commutative is an isomorphism of Weil groups. Indeed, the exactness of the sequences shows that φ is a group isomorphism of E on E', and $\varphi(E_U) = E'_U$ for all subgroups U of G. Hence φ induces an isomorphism

$$\varphi_U^c : E_U/E_U^c \to E'_U/E'^c_U.$$

We consider the cube:

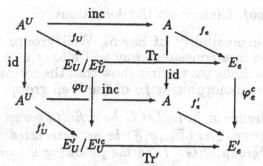

The top and bottom squares are commutative by **ISOW 1**. The back square is clearly commutative. The front face is commutative because the transfer is functorial. The square on the right is commutative because of the commutative diagram in Theorem 3.2. Hence the left square is commutative because the horizontal morphisms are injective.

Thus the study of a Weil isomorphism is reduced to the study of φ in the diagram. Such φ always exists since the group extensions have the same cohomology class. Uniqueness follows from the fact that $H^1(G, A) = 0$, using Theorem 1.3 of Chapter VIII, which was put there for the present purpose.

We already know that a class formation gives rise to others by restriction or deflation with respect to a normal subgroup. Similarly, a Weil group for (G, A) gives rise to Weil groups at intermediate levels as follows.

Theorem 3.3. *Let (G, A) be a class formation, and suppose G finite. Let (E_G, g_G, F_G) be the corresponding Weil group, where $\mathfrak{F}_G$ is the family of isomorphisms $\{f_U\}$ for subgroups U of G. Let V be a subgroup of G. Let*

$E_V = g_G^{-1}(V)$ *and* $g_V =$ *restriction of g_G to V;*
$\mathfrak{F}_V =$ *subfamily of $\mathfrak{F}_G$ consisting of those f_U such that $U \subset V$.*

Then:

(i) *$(E_V, g_V, \mathfrak{F}_V)$ is a Weil group associated to (V, A).*
(ii) *If V is normal in G, then $(E_G/E_V^c, \bar{g}_G, \bar{\mathfrak{F}}_G)$ is a Weil group associated with $(G/V, A^V)$, the family $\bar{\mathfrak{F}}_G$ consisting of the isomorphism*

$$\bar{f}_U : A^U \to E_U/E_U^c \approx (E_U/E_V^c)/(E_U^c/E_V^c),$$

where U ranges over the subgroups of G containing V.

Proof. Clear from the definitions.

The possibility of having Weil groups associated with factor groups in a consistent way will allows us to take an inverse limit. Before doing so, we first show that the reciprocity maps are induced by the isomorphisms f_U of the Weil group.

Theorem 3.4. *Let G be a finite group and (G, A) a class formation. Let $(E_G, g, \mathfrak{F})$ be an associated Weil group. Let V be a subgroup of G. Then the following diagram is commutative.*

$$
\begin{array}{ccc}
A^V & \xrightarrow{\ f_V\ } & E_V/E_V^c \\
\Big\downarrow{\scriptstyle S_G^V} & & \Big\downarrow{\scriptstyle \mathrm{inc}} \\
A^G & \xrightarrow[\ f_G\ ]{} & E_G/E_G^c
\end{array}
$$

Proof. Since we have not assumed that V is normal in G, we have to reduce the proof to this special case by means of a cube:

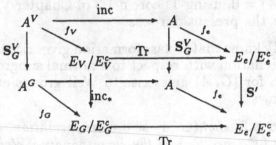

The vertical arrow $\mathbf{S}_G^V$ on the back face is defined by means of representatives of cosets of V in G. The front vertical arrow $\mathbf{S}'$ is defined to make the right face commutative. In other words, we lift these representatives in E_G be means of g^{-1}. Thus if $G = \bigcup \sigma_i V$ we choose $u_i \in E_G$ such that

$$g(\sigma_i) = u_i$$

and we define

$$g'(x) = \prod_i x^{u_i} (\mathrm{mod}\ E_e^c).$$

We note that $E_G = \bigcup u_i E_V$, in other words that the u_i represent the cosets of E_V in E_G. Then the front face is commutative, that is

$$\mathbf{S}'(\mathrm{Tr}'(u)) = \mathrm{Tr}(\mathrm{inc}_*(u)) \text{ for } u \in E_V/E_V^c,$$

immediately from the definition of the transfer. It then follows that the left face is commutative, thus finishing the proof.

Corollary 3.5. *Let G be finite and (G, A) a class formation. Let $(E_G, g, \mathfrak{F})$ be an associated Weil group. If $U \subset V$ are subgroups of G, then f_U and f_V induce isomorphisms:*

$$A^V / S_V^U(A^U) \approx E_V / E_U E_V^c \text{ and } (S_V^U)^{-1}(e) \approx (E_U \cap E_V^c) E_U^c.$$

If U is normal in V, then the first isomorphism is the reciprocity mapping, taking into account the isomorphism $E_V / E_U \approx V / U$.

Note that Corollary 3.5 is essentially the same result as Theorem 2.8. The proof of Corollary 3.5 is done by expliciting the transfer in terms of the Nakayama map, and the details are left to the reader.

In practice, in the context of class field theory, the group A has a topology (idele classes globally or multiplicative group of a local field locally). We shall now sketch the procedure which axiomatizes this topology, and allows us to take an inverse limit of Weil groups.

Let G be a group of Galois type and $A \in \mathrm{Galm}(G)$. We say that A is a **topological Galois module** if the following conditions are satisfied:

TOP 1. Each A^U (for U open subgoup of G) is a topological group, and if $U \subset V$, the topology of A^V is induced by the topology of A^U.

TOP 2. The group G acts continuously on A and for each $\sigma \in G$, the natural map $A^U \to A^{U[\sigma]}$ is a topological isomorphism.

Note that if $U \subset V$, it follows that the trace $S_V^U : A^U \to A^V$ is continuous.

Let G be of Galois type and $A \in \mathrm{Galm}(G)$ topological. If (G, A) is a class formation, we then say it is a **topological class formation**. By a **Weil group associated to such a topological class formation**, we mean a triplet $(E_G, g, \mathfrak{F})$ consisting of a topological group E_G, a morphism $g : E_G \to G$ in the category of topological groups (i.e. a continuous homomorphism) whose image is dense in G (so that for each open subgroups $V \supset U$ with U normal in G we have an isomorphism $E_V / E_U \approx V / U$), and a family of topological isomorphisms

$$f_U : A^U \to E_U / E_U^c$$

(where E_U^c is the closure of the commutator group), satisfying the following four axioms.

WT 1. For each pair of open subgroups $U \subset V$ of G, the following diagram is commutative:

$$
\begin{array}{ccc}
A^U & \xrightarrow{\ f_U\ } & E_U/E_U^c \\
{\scriptsize \text{inc}} \downarrow & & \downarrow {\scriptsize \text{Tr}} \\
A^V & \xrightarrow[\ f_V\ ]{} & E_V/E_V^c
\end{array}
$$

Note that the transfer on the right makes sense, because it extends continuously to the closure of the commutator subgroups.

WT 2. Let $x \in E_G$ be such that $\sigma = g(x)$. Then for all open subgroups U of G the following diagram is commutative:

$$
\begin{array}{ccc}
A^U & \xrightarrow{\ f_U\ } & E_U/E_U^c \\
{\scriptsize [\sigma]} \downarrow & & \downarrow {\scriptsize [x]} \\
A^{U[\sigma]} & \xrightarrow[\ f_{U[\sigma]}\ ]{} & E_{U[x]}/E_{U[x]}^c
\end{array}
$$

WT 3. If $U \subset V$ are open subgroups of G with U normal in V, then the class of the extension

$$
0 \to A^U \approx E_U/E_U^c \to E_V/E_U^c \to E_V/E_U \approx V/U \to 0
$$

is the fundamental class of $H^2(V/U, A^U)$.

WT 4. The intersection $\bigcap E_U^c$ taken over all open subgroups U of G is the unit element e of G.

To prove the existence of a topological Weil group, we shall need two sufficient conditions as follows.

WT 5. The trace $S_V^U : A^U \to A^V$ is an open morphism for each pair of open subgroups $U \subset V$ of G.

WT 6. The factor group A^U/A^V is compact.

Then there exists a topological Weil group associated to the formation.

Theorem 3.6. *Let G be a group of Galois type, $A \in \mathrm{Galm}(G)$, and (G, A) a topological class formation satisfying* **WT 5** *and* **WT 6**.

Proof. The proof is essentialy routine, except for the following remarks. In the uniqueness theorem for Weil groups when G is finite, we know that the isomorphism φ is determined only up to an inner automorphism by an element of $A = E_e$. When we want to take an inverse limit, we need to find a compatible system of Weil groups for each open U, so the topological A^U intervene at this point. The compactness hypothesis is sufficient to allow us to find a coherent system of Weil groups for pairs $(G/U, A^U)$ when U ranges over the family of open normal subgroups of G. The details are now left to the reader.

CHAPTER X

Applications of Galois Cohomology in Algebraic Geometry

by

John Tate

Notes by
Serge Lang

1959

Let k be a field and G_k the Galois group of its algebraic closure (or separable closure). It is compact, totally disconnected, and inverse limit of its factor groups by normal open subgroups which are of finite index, and are the Galois groups of finite extension.

We use the category of Galois modules (discrete topology on A, continuous operation by G) and a cohomological functor H_G such that $H(G, A)$ is the limit of $H(G/U, A^U)$ where U is open normal in G.

The Galois modules $\mathrm{Galm}(G)$ contains the subcategory of the torsion (for $\mathbf{Z}$) modules $\mathrm{Galm_{tor}}(G)$. Recall that G has **cohomological dimension** $\leq n$ if $H^r(G, A) = 0$ for all $r > n$ and all $A \in \mathrm{Galm_{tor}}(G)$, and that G has **strict cohomological dimen-**

sion $\leq n$ if $H^r(G,A) = 0$ for $r > n$ and all $A \in \text{Galm}(G)$.

We shall use the **tower theorem** that if N is a closed normal subgroup of G, then $\text{cd}(G) \leq \text{cd}(G/N) + \text{cd}(N)$, as in Chapter VII, Theorem 5.1. If a field k has trivial Brauer group, i.e.

$$H^2(G_E, \Omega^*) = 0 \qquad (\text{all } E/k \text{ finite})$$

where $\Omega = k_s$ (separable closure) then $\text{cd}(G_k) \leq 1$. By definition, a $\mathfrak{p}$-adic field is a finite extension of $\mathbf{Q}_p$. The maximal unramified extension of a $\mathfrak{p}$-adic field is cyclic and thus of $\text{cd} \leq 1$. Hence:

If k is a $\mathfrak{p}$-adic field, then $\text{cd}(G_k) \leq 2$.

This will be strengthened later to $\text{scd}(G_k) \leq 2$ (Theorem 2.3).

§1. Torsion-free modules

We use principally the dual of Nakayama, namely: Let G be a finite group, (G, A) a class formation, and M finitely generated torsion free (over $\mathbf{Z}$, and so $\mathbf{Z}$-free). Then

$$\mathbf{H}^r(G, \text{Hom}(M, A)) \times \mathbf{H}^{2-r}(G, M) \to H^2(G, A)$$

is a dual pairing, (i.e. puts the two groups in exact duality).

We suppose k is $\mathfrak{p}$-adic, and let Ω be its algebraic closure. We know G_k has $\text{cd}(G_k) \leq 2$ by the tower theorem. We shall eventually show $\text{scd}(G_k) \leq 2$.

Let $X = \text{Hom}(M, \Omega^*)$. Then X is isomorphic to a product of Ω^* as $\mathbf{Z}$-modules, their number being the rank of M, and we can define the operation of G_k on X in the natural manner, so that $X \in \text{Galm}(G_k)$.

By the existence theorem of local class field theory for L ranging over the finite extensions of k, the groups

$$N_{L/k} X_L$$

are cofinal with the groups nX_k, if we write $X_L = X^{G_L}$ and similarly $M_L = M^{G_L}$. This is clear since there exists a finite extension K of k which we may take Galois, such that $M^{G_K} = M$. Then for $L \supset K$, our statement is merely local class field theory's existence theorem, and then we use the norm $N_{L/K}$ to conclude the proof (transitivity of the norm).

We shall keep K fixed with the property that $M_K = M$. We wish to analyze the cohomology of X and M with respect to G_k.

Proposition 1.1. $H^r(G_k, X) = 0$ *for* $r > 2$.

Proof. X is divisible and we use cd ≤ 2, with the exact sequence

$$0 \to X_{\text{tor}} \to X \to X/X_{\text{tor}} \to 0,$$

where X_{tor} is the torsion part of X.

Theorem 1.2. *Induced by the pairing*

$$X \times M \to \Omega^*$$

we have the pairings

P0. $H^2(G_k, X) \times H^0(G_k, M) \to H^2(G_k, \Omega^*) = \mathbf{Q}/\mathbf{Z}$

P1. $H^1(G_k, X) \times H^1(G_k, M) \to \mathbf{Q}/\mathbf{Z}$

P2. $H^0(G_k, X) \times H^2(G_k, M) \to \mathbf{Q}/\mathbf{Z}$

In **P0**, $H^2(G_k, X) = M_k^\wedge =$ *by definition* $\mathrm{Hom}(M_k, \mathbf{Q}/\mathbf{Z})$.
In **P1**, *the two groups are finite, and the pairing is dual.*
In **P2**, $H^2(G_k, M)$ *is the torsion submodule of* $\mathrm{Hom}(X_k, \mathbf{Q}/\mathbf{Z})$, *i.e.*

$$H^2(G_k, M) = (X_k^\wedge)_{\text{tor}}.$$

Proof. The pairing in each case is induced by inflation in a finite layer $L \supset K \supset k$. In **P0**, the right hand kernel wil be the intersection of $N_{L/k} M_L$ taken for all $L \supset K$, and this is merely $[L : K] N_{K/k} M_K$ which shrinks to 0. The kernel on the left is obviously 0.

In **P1**, the inflation-restriction sequence together with Hilbert's Theorem 90 shows that

$$\mathrm{inf} : H^1(G_{K/k}, X_K) \to H^1(G_{L/k}, X_L)$$

is an isomorphism, and the trivial action of $G_{L/K}$ on M shows similarly that

$$\mathrm{inf} : H^1(G_{K/k}, M) \to H^1(G_{L/k}, M)$$

is an ismorphism. It follows that both groups are finite, and the duality in the limit is merely the duality in any finite layer $L \supset K$.

In **P3**, we dualize the argument of **P0**, and note that $H^2(G_k, M)$ will produce characters on X_k which are of finite period, i.e., which are trivial on some nX_k for some integer n. Otherwise, nothing is changed from the formalism of **P0**.

§2. Finite modules

The field k is again p-adic and we let A be a finite abelian group in $\mathrm{Galm}(G_k)$. Let $B = \mathrm{Hom}(A, \Omega^*)$. Then A and B have the same order, and $B \in \mathrm{Galm}(G_k)$. Since Ω^* contains all roots of unity, $B = \hat{A}$, and $A = \hat{B}$ once an identification between these roots of unity and $\mathbf{Q}/\mathbf{Z}$ has been made.

Let M be finitely generated torsion free and in $\mathrm{Galm}(G_k)$ such that we have an exact sequence in $\mathrm{Galm}(G_k)$:

$$0 \to N \to M \to A \to 0.$$

Since Ω^* is divisible, i.e. injective, we have an exact sequence

$$0 \leftarrow Y \leftarrow X \leftarrow B \leftarrow 0$$

where $X = \mathrm{Hom}(M, \Omega^*)$ and $Y = \mathrm{Hom}(N, \Omega^*)$.

By the theory of cup products, we shall have two dual sequences

(**B**) $H^1(B) \to H^1(X) \to H^1(Y) \to H^2(B) \to H^2(X) \to H^2(Y)$

(**A**) $H^1(A) \leftarrow H^1(M) \leftarrow H^1(N) \leftarrow H^0(A) \leftarrow H^0(M) \leftarrow H^0(N)$

with $H = H_{G_k}$. If one applies $\mathrm{Hom}(\cdot, \mathbf{Q}/\mathbf{Z})$ to sequence (**A**) we obtain a morphism of sequence (**B**) into $\mathrm{Hom}((\mathbf{A}), \mathbf{Q}/\mathbf{Z})$. The 5-lemma gives:

Theorem 2.1. *The cup product induced by $A \times B \to \Omega^*$ gives an exact duality*

$$H^2(G_k, B) \times H^0(G_k, A) \to H^2(G, \Omega^*).$$

Theorem 2.2. *With A again finite in* $\mathrm{Galm}(G_k)$ *and* $B = \mathrm{Hom}(A, \Omega^*)$ *the cup product*

$$H^1(G_k, A) \times H^1(G_k, B) \to H^2(G_k, \Omega^*)$$

gives an exact duality between the H^1*, both of which are finite groups.*

Proof. Let us first show they are finite groups. By the inflation-restriction sequence, it suffices to show that for L finite over k and suitably large, $H^1(G_L, A)$ is finite. Take L such that G_L operates trivially on A. Then $H^1(G_L, A) = \mathrm{cont}\ \mathrm{Hom}(G_L, A)$ and it is known from local class field theory or otherwise, that $G_L/G_L^n G_L^c$ is finite.

Now we have two sequences

(B) $H^0(B) \to H^0(X) \to H^0(Y) \to H^1(B) \to H^1(X) \to H^1(Y)$

(A) $H^2(A) \leftarrow H^2(M) \leftarrow H^2(N) \leftarrow H^1(A) \leftarrow H^1(M) \leftarrow H^1(N).$

We get a morphism from sequence **(A)** into the torsion part of $\mathrm{Hom}((\mathbf{B}), \mathbf{Q}/\mathbf{Z}))$, and the 5-lemma gives the desired result.

Next let n be a large integer, and let us look at the sequences above with $A = M/nM$ and $N = M$. We have the left part of our sequences

$$0 \to H^0(B) \to H^0(X)$$
$$H^3(M) \leftarrow H^3(M) \leftarrow H^2(A) \leftarrow H^2(M)$$

I contend that $H^2(M) \to H^2(A)$ is surjective, because every character of $H^0(B)$ is the restriction of some character of $H^0(X)$ since $B_k \cap nX_k = 0$ for n large. Hence the map

$$n : H^3(M) \to H^3(M)$$

is injective for all n large, and since we deal with torsion groups, they must be 0. This is true for every M torsion free finitely generated, in $\mathrm{Galm}(G_k)$. Looking at the exact sequence factoring out the torsion part, and using $\mathrm{cd} \le 2$, we see that in fact:

Theorem 2.3. *We have* $\mathrm{scd}(G_k) \leq 2$, *i.e.* $H_{G_k}^r = 0$ *for* $r \geq 3$ *and any object in* $\mathrm{Galm}(G_k)$.

We let χ be the (multiplicative) Euler characteristic. cf. *Algebra*, Chapter XX, §3.

Theorem 2.4. *Let* k *be* $\mathfrak{p}$*-adic, and* A *finite* G_k *Galois module. Let* $B = \mathrm{Hom}(A, \Omega^*)$. *Then* $\chi(G_k, A) = \|A\|_k$.

Proof. The Euler characteristic χ is multiplicative, so can assume A simple, and thus a vector space over $\mathbb{Z}/\ell\mathbb{Z}$ for some prime ℓ. We let $A_K = A^{G_K}$. For each Galois K/k, either $A_k = 0$ or $A_k = A$ by simplicity.

Case 1. $A_k = A$, so G_k operates trivially, so order of A is prime ℓ. Then

$$h^0 = \ell, \quad h^1 = (k^* : k^{*\ell}), \quad h^2(A) = h^0(B) = (k_\ell^* : 1).$$

So the formula checks.

Case 2. $B_k = B$. The situation is dual, and checks also.

Case 3. $A_k = 0$ and $B_k = 0$. Then

$$\chi(G_k, A) = 1/h^1(G_k, A).$$

Let K be maximal tamely ramified over k. Then G_K is a p-group. If $A_K = 0 = A^{G_K}$ then $\ell \neq p$ (otherwise G_K must operate trivially). Hence $H^1(G_K, A) = 0$. The inflation restriction sequence of G_k, G_K and $G_{K/k}$ shows $H^1(G_k, A) = 0$, so we are done.

Assume now $A_K \neq 0$, so $A_K = A$. Let L_0 be the smallest field containing k such that $A_{L_0} = A$. Then L_0 is normal over k, and cannot contain a subgroup of ℓ-power order, otherwise stuff left fixed would be a submodule $\neq 0$, so all of A, contradicting L_0 smallest. In particular, the ramification index of L_0/k is prime to ℓ.

Adjoin ℓ-th roots of unity to L_0 to get L. Then L has the same properties, and in particular, the ramification index of L/k is prime to ℓ.

Let T be the inertia field. Then the order of $G_{L/T}$ is prime to ℓ.

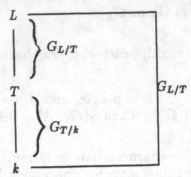

Hence $H^r(G_{L/T}, A) = 0$ for all $r > 0$. By spectral sequence, we conclude $H^r(G_{L/k}, A_L) = H^r(G_{T/k}, A_T)$ all $r > 0$. But $G_{T/k}$ is cyclic, A_T is finite, hence $H^1(G_{L/k}, A_L)$ and $H^2(G_{L/k}, A_L)$ have the same number of elements. In the exact sequence

$$0 \longrightarrow H^1(G_{L/k}, A_L) \longrightarrow H^1(G_k, A) \longrightarrow H^1(G_L, A)^{G_{L/k}} \longrightarrow H^2(G_{L/k}, A) \longrightarrow 0$$

we get 0 on the right, because $H^2(G_k, A)$ is dual to $H^0(G_k, B) = B_k = 0$. We can replace $H^2(G_{L/k}, A)$ by $H^1(G_{L/k}, A)$ as far as the number of elements is concerned, and then the hexagon theorem of the Herbrand quotient shows

$$h^1(G_k, A) = \text{ order } H^1(G_L, A)^{G_{L/k}}.$$

Since G_L operates trivially on A, the H^1 is simply the homs of G_L into A, and thus we have to compute the order of

$$\text{Hom}_{G_{L/k}}(G_L, A).$$

Such homs have to vanish on G_L^ℓ and on the commutator group, so if we let G_L' be the abelianized group, then $G_L^*/G_L^{*\ell}$. But this is $G_{L/k}$-isomorphic to $L^*/L^{*\ell}$, by local class field theory. So we have to compute the order of

$$\text{Hom}_{G_{L/k}}(L^*/L^{*\ell}, A).$$

If $\ell \neq p$, then $L^*/L^{*\ell}$ is $G_{L/k}$-isomorphic to $\mathbf{Z}/\ell\mathbf{Z} \times \mu_\ell$ where μ_ℓ is the group of ℓ-th roots of unity. Also, $G_{L/k}$ has trivial action on $\mathbf{Z}/\ell\mathbf{Z}$. So no hom can come from that since $A_k = 0$. As for $\mathrm{Hom}_{G_{L/k}}$ of μ_ℓ, if f is such, then for all $\sigma \in G_{L/k}$,

$$f(\sigma\zeta) = \sigma f(\zeta).$$

But $\sigma\zeta = \zeta^\nu$ for some ν, so $a = f(\zeta)$ generates a submodule of order ℓ, which must be all of A, so its inverse gives an element of B_k, contradicting $B_k = 0$. Hence all $G_{L/k}$-homs are 0, so Q.E.D.

If $\ell = p$, we must show the number of such homs is $1/\|A\|_k$. But according to Iwasawa,

$$L^*/L^{*p} = \mathbf{Z}/p\mathbf{Z} \times \mu_p \times \mathbf{Z}_p(G_{L/k})^m.$$

Using some standard facts of modular representations, we are done.

§3. The Tate pairing

Let V be a complete normal variety defined over a field k such that any finite set of points can be represented on an affine k-open subset of V. We denote by $A = A(V)$ its Albanese variety, defined over k, and by $B = B(V)$ its Picard variety also defined over k. Let $D_a(V)$ and $D_\ell(V)$ be the groups of divisors algebraically equivalent to 0, resp. linearly equivalent to 0. We have the Picard group $D_a(V)/D_\ell(V)$ and an isomorphism between this group and B, induced by a Poincaré divisor D on the product $V \times B$, and rational over k.

For each finite set of simple points S on V we denote by $\mathrm{Pic}_S(V)$ the factor group $D_{a,S}/D_{\ell,S}^{(1)}$ where $D_{a,S}$ consists of divisors algebraically equivalent to 0 whose support does not meet S, and $D_{\ell,S}^{(1)}$ is the subgroup of $D_{a,S}$ consisting of the divisors of functions f such that $f(P) = 1$ for all points P in S. Then there are canonical surjective homomorphisms

$$\mathrm{Pic}_{S'}(V) \to \mathrm{Pic}_S(V) \to \mathrm{Pic}(V)$$

whenever $S' \supset S$.

Actually we may work rationally over a finite extension K of k which in the applications will be Galois, and with obvious definitions, we form

$$\mathrm{Pic}_{S,K}(V) = D_{a,S,K}/D_{\ell,S,K}^{(1)}$$

the index K indicating rationality over K.

We may form the inverse limit inv $\lim_S \mathrm{Pic}_{S,K}(V)$. For our purposes we assume merely that we have a group $C_{a,K}$ together with a coherent set of surjective homomorphisms

$$\varphi_S : C_{a,K} \longrightarrow \mathrm{Pic}_{S,K}(V)$$

thus defining a homomorphism φ (their limit) whose kernel is denoted by U_K. We have therefore the exact sequence

$$\textbf{(1)} \qquad 0 \to U_K \to C_{a,K} \xrightarrow{j} B(K) \to 0.$$

We assume throughout that a divisor class (for all our equivalences) which is fixed under all elements of G_K contains a divisor rational over K. Similarly, we shall assume throughout that the sequence

$$\textbf{(2)} \qquad 0 \to Z_{\alpha,K} \to Z_{0,K} \to A(K) \to 0$$

is exact, (where Z_0 are the 0-cycles of degree 0, and Z_α is the kernel of Albanese), for each finite extension K of k.

Relative to our exact sequences, we shall now define a Tate pairing.

Since $C_{a,K}$ is essentially a projective limit, we shall use the exact sequence

$$0 \to D^1_{\ell,S,K}/D^{(1)}_{S,K} \to \mathrm{Pic}_{S,K} \xrightarrow{j} B(k) \to 0$$

because if $S' \supset S$, then we have a commutative and exact diagram

$$
\begin{array}{ccccccccc}
0 & \longrightarrow & D_{\ell,S',K}/D^{(1)}_{\ell,S',K} & \longrightarrow & \mathrm{Pic}_{S',K} & \longrightarrow & B(k) & \longrightarrow & 0 \\
& & \downarrow & & \downarrow & & \downarrow \text{id} & & \\
0 & \longrightarrow & D_{\ell,S,K}/D^{(1)}_{\ell,S,K} & \longrightarrow & \mathrm{Pic}_{S,K} & \longrightarrow & B(k) & \longrightarrow & 0 \\
& & & & \downarrow & & & & \\
& & & & 0 & & & &
\end{array}
$$

Now we wish to define a pairing

$$Z_{0,K} \times U_K \to K^*.$$

Let $u \in U_K$, and $\mathfrak{a} \in Z_{0,K}$. Write

$$\mathfrak{a} = \sum n_Q Q$$

where the Q are distinct algebraic points. Let S be a finite set of
points containing all those of $\mathfrak{a}$, and rational over K. Then u has
a representative in $\mathrm{Pic}_{S,K}$ and a further representative function f_S
defined over K, and defined at all points of $\mathfrak{a}$. We define

$$\langle \mathfrak{a}, u \rangle^{-1} = f_S(\mathfrak{a}) = \prod f_S(Q)^{n_Q}.$$

It is easily seen that $\langle \mathfrak{a}, u \rangle^{-1}$ does not depend on the choice of S
and f_S subject to the above conditions. It is then clear that this is
a bilinear pairing.

We define a further pairing

$$Z_{\alpha,K} \times C_{\mathfrak{a},K} \to K^*$$

as follows. Let $\gamma \in C_{\mathfrak{a},K}$, so $\gamma = \lim \gamma_S, \gamma_S \in D_{\mathfrak{a},S,K}/D_{\ell,S,K}^{(1)}$. Let
$\mathfrak{a} \in Z_{\alpha,K}$ and let S contain $\mathrm{supp}(\mathfrak{a})$. Let $\mathfrak{b}$ be a 0-cycle on B,
rational over K, corresponding to the point $b = j\gamma$. Let X_S be a
divisor on V rational over K, representing γ_S. Let D be a Poincaré
divisor whose support does not meet that of $S \times \mathfrak{b}$. We define:

$$\langle \mathfrak{a}, \gamma \rangle = \frac{[{}^t D(\mathfrak{b}) - X_S](\mathfrak{a})}{D(\mathfrak{a}, \mathfrak{b})}$$

observing that ${}^t D(\mathfrak{b}) - X_S$ is the divisor of a function whose support
does not meet S and is thus defined at $\mathfrak{a}$.

Using the reciprocity law of [La 57], see also [La 59], Chapter
VI, §4, Theorem 10, one verifies that this is independent of the
successive choice of $S, \mathfrak{b}, X_S$, and D subject to the above conditions.

One verifies finally that our pairings agree on $Z_\alpha \times U$.

Thus to summarize: we have exact sequences

$$0 \to Z_{\alpha,K}(V) \to Z_{0,K}(V) \to A(K) \to 0$$
$$0 \to U_K \to C_{\mathfrak{a},K}(V) \to B(K) \to 0$$

and we have a Tate pairing:

$$\langle \mathfrak{a}, \gamma \rangle = \frac{[^t D(\mathfrak{b}) - X_S](\mathfrak{a})}{D(\mathfrak{a}, \mathfrak{b})} \in K^*$$

where $\mathfrak{b} \in Z_0(B)$ maps on the point $b \in B$, the same point as $\gamma \in C_a(V)$, S is a finite set of points containing $\mathrm{supp}(\mathfrak{a}), X_S$ represents γ_S, and

$$\langle \mathfrak{a}, u \rangle = f_S(\mathfrak{a})^{-1}$$

where f_S is a function representing u. We take S so large that everything is defined.

Proposition 3.1. *The induced bilinear map on* (A_m, B_m) *coincides with* $e_m(a, b)$, *i.e. with* $^t D(m\mathfrak{b}, \mathfrak{a})/D(m\mathfrak{a}, \mathfrak{b})$.

Proof. Clear. We are using [La 57] and [La 59], Chapter VI.

The above statements refer to the Tate augmented product of Chapter V. The augmented product exists whenever one is given two exact sequences

$$0 \to A' \to A \xrightarrow{j} A'' \to 0$$

$$0 \to B' \to B \xrightarrow{j} B'' \to 0$$

an object C, two pairings $A \times B' \to C$ and $A' \times B \to C$ which agree on $A' \times B'$. Such an abstract situation induces an augmented product

$$H^r(A'') \times H^s(B'') \xrightarrow{\cup_a} H^{r+s+1}(C)$$

which may be defined in terms of cocycles as follows. If f'' and g'' are cocycles in A'' and B'' respectively, their augmented cup is represented by the cocycle

$$\delta f \cup g + (-1)^{\dim f} f \cup \delta g$$

where $jf = f''$ and $jg = g''$, i.e. f and g are cochains of A and B respectively pulled back from f'' and g''.

In dimensions $(0,1)$, the most important for what follows, we may make the Tate pairing explicit in the following manner. Let (b_σ) be a 1-cocycle representing an element $\beta \in H^1(G_{K/k}, B(K))$

and let $a \in A(K)$ represent $\alpha \in H^0(G_{K/k}, A(K))$. Let $\mathfrak{a} \in Z_{0,k}(V)$ belong to a, and let $\mathfrak{b}_\sigma$ be in $Z_{0,K}(B)$ and such that $S(\mathfrak{b}_\sigma) = b_\sigma$. Let D be a Poincaré divisor on $V \times B$ whose support does not meet $\mathfrak{a} \times \mathfrak{b}_\sigma$ for any σ. Then it is easily verified that putting $\mathfrak{b} = \mathfrak{b}_\sigma + \sigma \mathfrak{b}_\tau - \mathfrak{b}_{\sigma\tau}$, the cocycle

$$^t D(\mathfrak{b}, \mathfrak{a})$$

represents $\alpha \cup_{\text{aug}} \beta$.

§4. The $(0,1)$ duality for abelian varieties

We assume for the rest of this section that k is $\mathfrak{p}$-adic.

Theorem 4.1. *The augmented product of the Tate pairing described in Section 3 induces a duality between $H^0(G_k, A)$ and $H^1(G_k, B)$, with values in $H^2(G_k, \Omega) = \mathbf{Q}/\mathbf{Z}$.*

Proof. According to the general theory of the augmented cupping, we have for each integer $m > 0$,

$$
\begin{array}{ccccccccc}
0 & \to & A(k)/mA(k) & \to & H^1(G_k, A_m) & \to & H^1(G_k, A)_m & \to & 0 \\
& & \downarrow & & \downarrow & & \downarrow & & \\
0 & \to & (H^1(G_k, B)_m)^\wedge & \to & H^1(G_k, B_m)^\wedge & \to & (B(k)/mB(k))^\wedge & \to & 0
\end{array}
$$

a morphism of the first sequence into the second. Since the pairing between A_m and B_m is an exact duality, so is the pairing between their H^1 by Theorem 2.1. We wish to prove the end vertical arrows are isomorphisms, and for this we count. We have:

$$(A(k) : mA(k)) \le h^1(B)_m \qquad (B(k) : mB(k)) \le h^1(A)_m$$

$$h^1(A)_m (A(k) : mA(k)) = h^1(A_m)$$

$$h^1(B)_m (B(k) : mB(k)) = h^1(B_m)$$

$$\frac{(A(k) : mA(k))}{(A(k)_m : 0)} = \|m\|_k^{-r} = \frac{(B(k) : mB(k))}{(B(k)_m : 0)}$$

(by cyclic cohomology, trivial action)

$$\chi(A_m) = \|m^{2r}\|_k = \chi(B_m) \quad \text{and} \quad h^0(B_m) = h^2(A_m)$$

by duality. Putting everything together, we get equality in the first inequalities; this proves the desired isomorphism.

Theorem 4.2. *We have* $H^2(G_k, B) = 0$. *(This is special for abelian varieties, better than* scd ≤ 2.)

Proof. We have an exact sequence

$$0 \to H^1(B)/mH^1(B) \to H^2(B_m) \to H^2(B_m) \to H^2(B)_m \to 0.$$

But $H^2(B_m)$ is dual to $H^0(A_m)$ and in particular has the same number of elements. Also, $H^1(B)$ being dual to $H^0(A)$, we see that $H^1(B)/mH^1(B)$ is dual to $H^0(A)_m = A(k)_m$, which is also $H^0(A_m)$. Hence the two terms on the left have the same number of elements, since $H^2(B)_m = 0$ for all m, so 0 since it is torsion group.

Now we have the duality for H^1, H^0 in finite layers.

Theorem 4.3. *Let K/k be finite Galois with group $G = G_{K/k}$. Then the pairing*

$$H^0(G, A(K)) \times H^1(G, B(K)) \to H^2(G, K^*)$$

is a duality.

Proof. This follows from the abstract fact that restriction is dual to the transfer valid for any Tate pairing and the induced augmented cupping.

If one uses the inflation-restriction sequence, together with the commutativity derived abstractly for d_2, and Theorem 4.2, we get the following $\pm$ commutative diagram, putting $U = G_K$, and $G = G_{K/k}$,

$$
\begin{array}{ccccc}
H^1(U,A)^{G/U} \times (B^U)_{G/U} & \xrightarrow{U_{aug}} & H^2(U,\Omega^*)_{G/U} & \xrightarrow{\text{tr}} & H^2(G,\Omega^*) \\
\downarrow{d_2} \qquad \uparrow{\text{inc}} & & & & \downarrow{\text{id.}} \\
H^2(G/U,A^U) \times H^1(G/U,B^U) & \xrightarrow{U_{aug}} & H^2(G/U,\Omega^{*U}) & \xrightarrow{\text{inf}} & H^2(G,\Omega^*)
\end{array}
$$

Identifying $H^2(G, \Omega^*)$ with $\mathbf{Q}/\mathbf{Z}$ we get the duality between H^2 and $\mathbf{H}^{-1}$:

Identifying $H^2(G, \Omega^*)$ with $\mathbf{Q}/\mathbf{Z}$ we get the duality between H^2 and $\mathbf{H}^{-1}$:

Theorem 4.4. *If K/k is a finite Galois extension with group G, then the augmented cupping*

$$\mathbf{H}^2(G, A(K)) \times \mathbf{H}^{-1}(G, B(K)) \to \mathbf{H}^2(G, K^*)$$

is a perfect duality.

§5. The full duality

We wish to show how the following theorem essentially follows from the $(0, 1)$ duality without any further use of arithmetic, only from abstract commutative diagrams.

Theorem 5.1. *Let k be a p-adic field, A and B an abelian variety and its Picard variety defined over k, and consider the Tate pairing described in §3. Then the augmented cupping*

$$\mathbf{H}^{1-r}(G_{K/k}, A(K)) \times \mathbf{H}^r(G_{K/k}, B(K)) \to \mathbf{H}^2(G_{K/k}, K^*)$$

puts the two groups (which are finite) in exact duality.

(Of course, the right hand H^2 is $(\mathbf{Q}/\mathbf{Z})_n$, where $n = (G_{K/k} : e)$.)

As usual, the $\wedge$ means Hom into $\mathbf{Q}/\mathbf{Z}$.

Put $G_K = U$ and $G = G_k$. We have a compact discrete duality

$$A^U \times H^1(U, B) \to H^2(U, \Omega^*) = \mathbf{Q}/\mathbf{Z}$$

and we know from this that A^U is isomorphic to $H^1(U, B)^\wedge$ as a G/U-module. Hence the commutative diagram say for $r \geq 3$:

$$
\begin{array}{ccccc}
\mathbf{H}^{1-r}(G/U, H^1(U,B)^\wedge) & \times & \mathbf{H}^{r-2}(G/U, H^1(U,B)) & \xrightarrow{\ \cup\ } & \mathbf{H}^{-1}(G/U, \mathbf{Q}/\mathbf{Z}) \\
\downarrow & & \downarrow \text{id} & & \downarrow \text{id} \\
\mathbf{H}^{1-r}(G/U, A^U) & \times & \mathbf{H}^{r-2}(G/U, H^1(U,B)) & \xrightarrow[\cup]{\ } & \mathbf{H}^{-1}(G/U, \mathbf{Q}/\mathbf{Z}).
\end{array}
$$

The top line comes from the cup product duality theorem, and the arrow on the left is an ismorphism, as described above.

Of course, we have $H^{-1}(G/U, \mathbf{Q}/\mathbf{Z}) = (\mathbf{Q}/\mathbf{Z})_n$ if n is the order of G/U. Note also that the $\mathbf{Q}/\mathbf{Z}$ in the lower right stands for $H^2(U, \Omega^*)$, because of the invariant isomorphism.

Now from the spectral sequence and Theorem 4.2 to the effect that H^2 of an abelian variety is trivial over a $\mathfrak{p}$-adic field, we get an isomorphism

$$d_2 : H^{r-2}(G/U, H^1(U, B)) \to H^r(G/U, H^0(U, B))$$

and we use another abstract diagram:

$$
\begin{array}{ccc}
\mathbf{H}^{1-r}(G/U,A^U) \times \mathbf{H}^{r-2}(G/U,H^1(U,B)) & \overset{\cup}{\longrightarrow} & \mathbf{H}^{-1}(G/U,H^2(U,\Omega^*)) \\
\Big\downarrow \text{id} & \Big\downarrow d_2 & \\
\mathbf{H}^{1-r}(G/U,A^U) \times \mathbf{H}^r(G/U,H^0(U,B))) & \underset{\cup_{\text{aug}}}{\longrightarrow} & \mathbf{H}^2(G/U,\Omega^{*U})
\end{array}
$$

In order to complete it to a commutative one, we complete the top line and the bottom one respectively as follows:

$$
\begin{array}{ccc}
\mathbf{H}^{-1}(G/U,H^2(U,\Omega^*)) \underset{\text{inc}}{\longrightarrow} H^2(U,\Omega^*)_{G/U} \overset{\text{tr}}{\longrightarrow} H^2(G,\Omega^*) \\
\Big\downarrow \text{id.} \\
\mathbf{H}^2(G/U,\Omega^{*U}) \underset{\text{inf}}{\longrightarrow} H^2(G,\Omega^*)
\end{array}
$$

and since the transfer and inflation perserve invariants, we see that our duality has been reduced as advertised.

We observe that we have the ordinary cup on the top line and the augmented cup on the bottom. The top one is relative to the $A^U, H^1(U, B)$ duality, derived previously.

§6. The Brauer group

We continue to work with a variety V defined over a $\mathfrak{p}$-adic field k. We assume V complete, non-singular in codimension 1, and such that any finite set of points can be represented on an affine k-open subset of V.

We let $G = G_k$ and all cohomology groups in this section will be taken relative to G. We observe that the function field $\Omega(V)$ has group G over $k(V)$, and we wish to look at its cohomology.

By Hilbert's Theorem 90 it is trivial in dimension 1, and hence we look at it in dimension 2: It is nothing but that part of the Brauer group over $k(V)$ which is split by a constant field extension. We make the following assumptions.

Assumption 1. *There exists a 0-cycle on V rational over k and of degree 1.*

Assumption 2. *Let NS denote the Néron-Severi group of V (it is finitely generated). Then the natural map*

$$\mathrm{Div}(V)^{G_k} \to NS^{G_k} = NS_k$$

of divisors rational over k into that part of Néron-Severi which is fixed under G_k is surjective, i.e. every class rational over k has a representative divisor rational over k.

These assumptions can be translated into cohomology, and it is actually in this latter form that we shall use them. This is done as follows.

To begin with, note that Assumption 1 guarantees that there is a canonical map of V into its Albanese variety defined over k (use the cycle to get an origin on the principal homogeneous space of Albanese). Hence by pull-back from Albanese, given a rational point b on the Picard variety, there is a divisor $X \in D_a(V)$ rational over k such that $Cl(X) = b$. In other words, the map

$$D_a(V)^{G_k} \to B(k)$$

is surjective. Now consider the exact sequence

$$H^0(\mathrm{Div}(V)) \to H^0(NS) \to H^1(D_a) \to H^1(\mathrm{Div}(V)).$$

Then $\mathrm{Div}(V)^{G_k}$ is a direct sum over $\mathbf{Z}$ of groups generated by the irreducible divisors, and putting together a divisor and its conjugates, we get

$$\mathrm{Div}(V)^{G_k} = \bigoplus_{\xi} \bigoplus_{X \in \xi} \mathbf{Z} \cdot X$$

where ξ ranges over the prime rational divisors of V over k and X ranges over its algebraic components. Now the inside sum is semilocal, and by semilocal theory we get $H^1(G_K, \mathbf{Z})$ where $K = k_X$ is the smallest field of definition of X. This is 0 because G_K is of Galois type and the cohomology comes from finite things. Thus:

Proposition 6.1. $H^1(\mathrm{Div}(V)) = 0$.

From this, one sees that our Assumption 1 is equivalent with the condition $H^1(D_a) = 0$.

Now looking at the other sequence

$$H^0(D_a) \to H^0(B) \to H^1(D_\ell) \to H^1(D_a)$$

we see that Assumption 2 is equivalent to $H^1(D_\ell) = 0$. Thus:

Assumptions 1 and 2 are equivalent with

$$H^1(D_a) = 0 \quad and \quad H^1(D_\ell) = 0.$$

Now we have two exact sequences

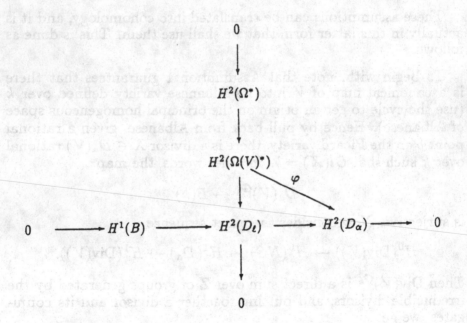

and from them we get a surjective map

$$\varphi : H^2(\Omega(V)^*) \to H^2(D_a) \to 0.$$

We *define* $H^2_u(\Omega(V)^*)$ to be its kernel, and call it the **unramified** part of the Brauer group $H^2(\Omega(V)^*)$. In view of the exact cross we get a map

$$H^2_u(\Omega(V)^*) \to H^1(B).$$

Thus

$$H^2(\Omega(V)^*)/H_u^2(\Omega(V)^*) \approx H^2(D_a)$$

$$H_u^2(\Omega(V)^*)/H^2(\Omega) \approx H^1(B) \approx \mathrm{Char}(A(k))$$

$$H^2(\Omega^*) \approx \mathbf{Q}/\mathbf{Z} = \mathrm{Char}(\mathbf{Z})$$

where Char means continuous character (or here equivalently character of finite order, or torsion part of $\hat{A}(k) = \mathrm{Hom}(A(k), \mathbf{Q}/\mathbf{Z})$).
This gives us a good description of our Brauer group, relative to the filtration

$$H^2(\Omega(V)^*) \supset H_u^2(\Omega(V)^*) \supset H^2(\Omega^*) \supset 0.$$

We wish to give a more concrete description of H_u^2 above, making explicit its connection with the Tate pairing. Relative to the sequence

$$0 \to A(k) \to Z_k/Z_{\alpha,k} \to \mathbf{Z} \to 0$$

and taking characters Char, we shall get a commutatiove exact diagram as follows:

$$
\begin{array}{ccccccccc}
0 & \longrightarrow & \mathrm{Char}(\mathbf{Z}) & \longrightarrow & \mathrm{Char}(Z_k/Z_{\alpha,k}) & \longrightarrow & \mathrm{Char}(A(k)) & \longrightarrow & 0 \\
 & & \uparrow & & \uparrow & & \uparrow & & \\
0 & \longrightarrow & H^2(\Omega^*) & \longrightarrow & H^2(\Omega(V)^*) & \longrightarrow & H^1(B) & \longrightarrow & 0
\end{array}
$$

The two end arrows are as we have just described them, and are isomorphisms. We must now define the middle arrow and prove commutativity.

Let $v \in H_u^2(\Omega(V)^*)$. For each prime rational 0-cycle $\mathfrak{p}$ of V over k we shall define its reduction mod $\mathfrak{p}, v_\mathfrak{p} \in H^2(\Omega^*)$ and then a character $Z_k/Z_{\alpha,k}$ by the formula

$$\theta_V(\mathfrak{a}) = \sum_\mathfrak{p} \nu_\mathfrak{p} \, \mathrm{inv}(v_\mathfrak{p})$$

whenever $\mathfrak{a}$ is a rational 0-cycle,

$$\mathfrak{a} = \sum \nu_\mathfrak{p} \cdot \mathfrak{p}, \; \nu_\mathfrak{p} \in \mathbf{Z}.$$

We shall prove that θ_V vanished on the kernel of Albanese, whence the character, and then we shall prove commutativity.

Definition of $v_\mathfrak{p}$. Let $(f_{\sigma,\tau})$ be a representative cocycle. By definition it splits in D_a, so that there is a divisor $X_\sigma \in D_a$ such that

$$(f_{\sigma,\tau}) = X_\sigma + \sigma X_\tau - X_{\sigma\tau}.$$

For each σ, choose a function g_σ such that $X_\sigma = (g_\sigma)$ at $\mathfrak{p}$. Put

$$f'_{\sigma,\tau} = f_{\sigma,\tau}/(\delta g)_{\sigma,\tau}$$

then $(f'_{\sigma,\tau}) = 0$ at $\mathfrak{p}$, i.e. $f'_{\sigma,\tau}$ is a unit at $\mathfrak{p}$. We now put

$$a_{\sigma,\tau} = \prod_{P \in \mathfrak{p}} f'_{\sigma,\tau}(P),$$

where $\{P\}$ ranges over the algebraic points into which $\mathfrak{p}$ splits. We contend that $(a_{\sigma,\tau})$ is a cocycle, and that its class does not depend on the choices made during its construction.

Let $(f^*_{\sigma,\tau})$ be another representative cocycle which is a unit at $\mathfrak{p}$, and obtained by the same process. Then

$$f^* = f' \cdot \delta g$$

with $(\delta g)_{\sigma,\tau} = g_\sigma g_\tau^\sigma / g_{\sigma\tau}$ a unit at $\mathfrak{p}$. Hence taking divisors,

$$(g_\sigma) + \sigma(g_\tau) - (g_{\sigma\tau}) = 0$$

at all points in $\operatorname{supp}(\mathfrak{p})$. Let this support be S, and let Div^S be the group of divisors passing through some point of $\mathfrak{p}$. Then $H^1(\operatorname{Div}^S) = 0$ by the same argument as in Proposition 6.1 (semilocal and $H^1(\mathbf{Z}) = 0$) and hence taking the image of (g_σ) in Div^S we conclude that there exists a divisor $X \in \operatorname{Div}^S$ such that $(g_\sigma) = \sigma X - X$. Let h be a function such that $X = (h)$ at $\mathfrak{p}$. Replace g_σ by $g_\sigma h^{1-\sigma}$. Then still $f^* = f' \cdot \delta g$ and now g_σ is a unit at $\mathfrak{p}$, for all σ. From this we get

$$\prod_{P \in \mathfrak{p}} (f^*/f')_{\sigma,\tau}(P) = \prod_{P \in \mathfrak{p}} (\delta g)_{\sigma,\tau}(P)$$

from which we see that it is the boundary of the 1-cochain

$$\prod_{P \in \mathfrak{p}} g_\sigma(P).$$

Thus we have proved our reduction mapping

$$v \mapsto v_{\mathfrak{p}}$$

well defined.

Now the image of v in $H^1(B)$ is by definition reprsented by the cocycle $Cl(X_\sigma) = b_\sigma$ (notation as in the above paragraph) and if $\mathfrak{a} \in Z_{0,k}$ then

$$\theta_v(\mathfrak{a}) = \pm S(\mathfrak{a}) \cup_{\text{aug}} \beta$$

where β is represented by the cocycle (b_σ).

Thus θ_v vanishes on the kernel of Albanese, and the right side of our diagam is commutative.

As for the left side, given $\omega \in H^2(\Omega^*)$, represented by a cocycle $(c_{\sigma,\tau})$, then $\omega_{\mathfrak{p}}$ is represented by

$$\prod_{P \in \mathfrak{p}} c_{\sigma,\tau} = c_{\sigma,\tau}^m$$

where $m = \deg(\mathfrak{p})$. Then

$$\theta_\omega(\mathfrak{a}) = \sum_{\mathfrak{p}} (\deg \mathfrak{p}) \nu_{\mathfrak{p}} \text{inv}(\omega)$$

$$= \deg(\mathfrak{a}) \cdot \text{inv}(\omega)$$

whence commutativity. This concludes the proof of the following theorem.

Theorem 6.2. *There is an isomorphism*

$$H_u^2(\Omega(V)^*) \approx \text{Char}(Z/Z_\alpha)_k$$

under the mapping θ_v and the diagram

$$
\begin{array}{ccccccccc}
0 & \longrightarrow & \text{Char}(Z) & \longrightarrow & \text{Char}(Z/Z_\alpha)_k & \longrightarrow & \text{Char}(A(k)) & \longrightarrow & 0 \\
 & & \uparrow & & \uparrow & & \uparrow & & \\
0 & \longrightarrow & H^2(\Omega^*) & \longrightarrow & H^2(\Omega(V)^*) & \longrightarrow & H^1(B)^\smallfrown & \longrightarrow & 0
\end{array}
$$

which is exact and commutative.

We conclude this section with a description of $H^2(D_a)$ also in terms of characters.

We have the exact sequence

$$0 \to D_a \to \mathrm{Div} \to NS \to 0$$

and hence

$$0 \to H^1(NS) \to H^2(D_a) \to H^2(\mathrm{Div}) \to H^2(NS) \to 0$$

the last 0 by scd ≤ 2.

Now $H^2(\mathrm{Div})$ is easy to describe since Div is essentially a direct sum. In fact,

$$\mathrm{Div}_k = \bigoplus_{\xi} \bigoplus_{X \in \xi} \mathbf{Z} \cdot X$$

where the sum is taken over all prime rational divisors ξ over k and all algebraic components X in ξ. Using the semilocal theory, we get

$$H^2(\mathrm{Div}) = \bigoplus_{\xi} H^2(G_{k_\xi}, \mathbf{Z}) = \bigoplus_{\xi} H^1(G_{k_\xi}, \mathbf{Q}/\mathbf{Z})$$

$$= \bigoplus_{\xi} \mathrm{Char}(k_\xi^*)$$

where k_ξ is the smallest field of definition of an algebraic point X in ξ and G_{k_ξ} is the Galois group over k_ξ. Thus we see that our H^2 is a direct sum of character groups.

The Case of Curves. If we assume that V has dimension 1, i.e. is a non singular curve, then this result simplifies considerably since $NS = \mathbf{Z}$ is infinite cyclic, and we have also

$$NS = NS_k = (NS)^{G_k}.$$

We have

$$H^1(NS) = 0, \quad H^2(NS) = H^2(\mathbf{Z}) = \mathrm{Char}(k^*)$$

and we get the commutative diagram

$$0 \longrightarrow H^2(D_a) \longrightarrow H^2(\mathrm{Div}) \longrightarrow H^2(\mathbf{Z}) \longrightarrow 0$$

$$\| \qquad\qquad\qquad \| \qquad\qquad\qquad \|$$

$$0 \longrightarrow \bigoplus{}^0 \mathrm{Char}(k_\xi^*) \longrightarrow \bigoplus \mathrm{Char}(k_\xi^*) \longrightarrow \mathrm{Char}(k^*) \longrightarrow 0$$

where the $\bigoplus^0$ on the left means those elements whose sum gives 0. The morphism on the lower right is given by the restriction of a character from k_ξ^* to k^* and the sum mapping. Thus an element of $H^2(\mathrm{Div})$ is given by a vector of characters $(\chi_{v,\mathfrak{p}})$ where $\mathfrak{p}$ ranges over the prime rational cycles of V, i.e. the ξ since cycles and divisors coincide.

We observe also that by Tsen's theorem, $H^2(\Omega(V)^*)$ is the full Brauer group over $k(V)$ since $\Omega(V)$ does not admit any division algebras of finite rank over itself.

Finally, we have slightly better information on H_u^2:

Proposition 6.3. *If V is a curve, then*

$$H_u^2(\Omega(V)^*) = \bigcap_{\mathfrak{p}} H_{\mathfrak{p}}^2(\Omega(V)^*)$$

where $H_{\mathfrak{p}}^2$ consists of those cohomology classes having a cocycle representative $(f_{\sigma,\tau})$ in the units at $\mathfrak{p}$.

We leave the proof as an exercise to the reader.

We shall discuss ideles for arbitrary varieties in the next section. Here, for curves, we take the usual definition, and we then have the same theorem as in class field theory.

Proposition 6.4. *Let V be a curve, and for each $\mathfrak{p}$ let $k(V)_\mathfrak{p}$ be the completion at the prime rational cycle $\mathfrak{p}$. Let $\mathrm{Br}(k(V))$ be the Brauer group over $k(V)$, i.e. $H^2(G_k, k(V)_s)$ where $k(V)_s$ is the separable ($=$ algebraic) closure of $k(V)$, and similarly for $\mathrm{Br}(k(V)_\mathfrak{p})$. Then the map*

$$\mathrm{Br}(k(V)) \to \prod_{\mathfrak{p}} \mathrm{Br}(k(V)_\mathfrak{p})$$

is injective.

One can give a proof based on the preceding discussion or by proving that $H^1(C_a) = 0$, just as in class field theory. We leave the details to the reader.

§7. Ideles and idele classes

Let k be a field and V a complete normal variety defined over k and such that any finite set of points can be represented on an affine k-open subset. By a cycle, we shall always mean a 0-cycle.

For each prime rational cycle $\mathfrak{p}$ over k on V we have the integers $\mathfrak{o}_\mathfrak{p}$, the units $U_\mathfrak{p}$ and maximal ideal $\mathfrak{m}_\mathfrak{p}$ in $k(V)$.

There are several candidates to play the role of ideles, and we shall describe here what would be a factor group of the classical ideles in the case of curves. We let

$$F_\mathfrak{p} = k(V)^*/(1 + \mathfrak{m}_\mathfrak{p}).$$

Then we let I_k be the subgroup of the Cartesian product of all the $F_\mathfrak{p}$ consisting of the vectors

$$\mathbf{f} = (\dots, f_\mathfrak{p}, \dots) \qquad f_\mathfrak{p} \in F_\mathfrak{p}$$

such that there exists a divisor X, rational over k, such that

$$X = (f_\mathfrak{p}) \text{ at } \mathfrak{p} \text{ for all } \mathfrak{p}.$$

(In the case of curves, this means unit almost everywhere.) We call this divisor X (obviously unique) the **divisor associated with the idele f**, and write $X = (\mathbf{f})$.

We have two subgroups $I_{a,k}$ and $I_{\ell,k}$ consisting of the ideles whose divisor is algebraically equivalent to 0 and linearly equivalent to 0 respectively.

Since every divisor is linearly equivalent to 0 at a simple point, we have an exact sequence

$$0 \to I_{\ell,k} \to I_{a,k} \to B(k) \to 0$$

where B is the Picard variety of V, defined over k.

As usual, we have an imbedding

$$K(V)^* \subset I_k$$

on the diagonal: if $f \in k(V)^*$, then f maps on $(\dots, f, f, f, \dots)$ (of course in the vector, it is the class of $f \mod 1 + \mathfrak{m}_\mathfrak{p}$).

We recall our Picard groups $\text{Pic}_S(V)$ associated with a finite set of points of V and here we assume that S is a finite set of prime rational cycles. We have $\text{Pic}_{S,k}(V) = D_{a,S,k}(V)/D_{\ell,S,k}^{(1)}$ where $D_{a,S,k}$ consists of the divisors on V algebraically equivalent to 0, not passing through any point of S, and rational over k, and $D_{\ell,S,k}^{(1)}$ consists of those which are linarly equivalent to 0, belonging to a function which takes the value 1 at all points of S, and is defined over k.

We contend that we have a surjective map

$$\varphi_S : I_{a,k} \longrightarrow \text{Pic}_{S,k}$$

for each S as follows. Given $\mathbf{f}$ in $I_{a,k}$ there exists $f \in k(V)^*$ such that we can write

$$\mathbf{f} = f f'_{\mathfrak{p}} \text{ with } \mathbf{f}' = 1 \in F_{\mathfrak{p}}$$

for all $\mathfrak{p} \in S$. This is easily proved by moving the divisor of $\mathbf{f}$ by a linear equivalence, and then using the Chinese remainder theorem in an affine ring of an affine open subset of V. We then put

$$\varphi_S(\mathbf{f}) = Cl_S((\mathbf{f}'))$$

where Cl_S is the equivalence class mod $D_{\ell,S,k}^{(1)}$. Our collection of maps φ_S is obviously consistent, and thus we can define a mapping

$$\varphi : I_{a,k} \longrightarrow \lim \text{Pic}_{S,k}(V).$$

For our purposes here, we denote by $C_{a,k}$ the image of φ in the limit and call it the group of **idele classes**. This is all right:

Contention. *The kernel of φ is $k(V)^*$.*

Proof. If $\mathbf{f}$ is in the kernel, then for all S there exists a function f_S such that

$$\mathbf{f} = \mathbf{f}_S f_S$$

where $\mathbf{f}_S$ is 1 in S, and $(f_S) = 0$. All f_S have the same divisor, namely $(\mathbf{f})$. Looking at one prime $\mathfrak{p}$ in S, we see that all f_S are equal to the same function f, and we see that $\mathbf{f}$ is simply the function f.

212

We have the **unit ideles** $I_{u,k}$ consisting of those ideles whose divisor is 0, the **idele classes** $C_k = I_k/K(V)^*$, and also the obvious subgroups of idele classes:

$$C_{a,k} = I_{a,k}/k(V)^*$$
$$C_{u,k} = k(V)^* I_{u,k}/k(V)^* = I_{u,k}/k^*.$$

We keep working under Assumptions 1 and 2, of course. In that case, if K is a finite Galois extension of k, we have the two fundamental exact sequences

(1) $\qquad 0 \to Z_{\alpha,K} \to Z_{0,K} \to A(K) \to 0$

(2) $\qquad 0 \to C_{u,K} \to C_{a,K} \to B(K) \to 0$

in the category of $G_{K/k}$-mpodules. For the limit, with respect to Ω one will of course take the injectve limit over all K.

From the definition, we see that

$$I_{u,k} = \prod_{\mathfrak{p}/k} k(\mathfrak{p})^*$$

where $k(\mathfrak{p})$ is the residue class field of the prime rational cycle $\mathfrak{p}$ over k, i.e. $k(\mathfrak{p}) = \mathfrak{o}_\mathfrak{p}/\mathfrak{m}_\mathfrak{p}$.

If K/k is finite Galois, then we write

$$I_{u,K} = \prod_{\mathfrak{P}/K} k(\mathfrak{P})^*$$

where $\mathfrak{P}$ ranges over the prime rational cycles over K.

§8. Idele class cohomology

Aside from the fundamental sequences (1) and (2), we have three sequences.

$$0 \to Z_{0,K} \to Z_K \to \mathbf{Z} \to 0$$
$$0 \to K^* \to I_{u,K} \to C_{u,K} \to 0$$
$$0 \to 0 \to K^* \to K^* \to 0$$

and pairings giving rise to cup products:

$$Z_K \times I_{u,K} \to K^*$$

defined in the obvious manner: Given $f \in I_{u,K}$ and a cycle

$$a = \sum_p \nu_p \cdot p,$$

the pairing is

$$(a, f) = \prod_p f_p^{\nu_p}.$$

It induces pairings

$$Z_{0,K} \times C_{u,K} \to K^*$$
$$\mathbf{Z} \times K^* \to K^*$$
$$Z_{0,K} \times K^* \to 0$$

and we get an exact commutative diagram from the cup product

$$
\begin{array}{ccccccc}
H^r(K^*) & \longrightarrow & H^r(I_u) & \longrightarrow & H^r(C_u) & \longrightarrow & H^{r+1}(K^*) \\
\downarrow \varphi_1 & & \downarrow \varphi_2 & & \downarrow \varphi_3 & & \downarrow \varphi_1 \\
H^{2-r}(\mathbf{Z})^\wedge & \longrightarrow & H^{2-r}(\mathbf{Z})^\wedge & \longrightarrow & H^{2-r}(Z_0)^\wedge & \longrightarrow & H^{1-r}(\mathbf{Z})^\wedge
\end{array}
$$

taking into account that

$$H^2(G_{K/k}, K^*) = (\mathbf{Q}/\mathbf{Z})_n$$

where $n = (G : e)$ the cup products taking their values in this H^2. Here, as in the next diagram, H is taken with respect to $G_{K/k}$, we omit the index K on the modules, and $r \in \mathbf{Z}$ so $H = H_{G_{K/k}}$ is the special functor.

From the exact sequence in the last section, we get

$$
\begin{array}{ccccccccc}
H^{r-1}(B) & \longrightarrow & H^r(C_u) & \longrightarrow & H^r(C_a) & \longrightarrow & H^r(B) & \longrightarrow & H^{r+1}(C_u) \\
\downarrow \varphi_4 & & \downarrow \varphi_3 & & \downarrow \varphi_5 & & \downarrow \varphi_4 & & \downarrow \varphi_3 \\
H^{2-r}(A)^\wedge & \longrightarrow & H^{2-r}(Z)_0)^\wedge & \longrightarrow & H^{2-r}(Z_a)^\wedge & \longrightarrow & H^{1-r}(A)^\wedge & \longrightarrow & H^{1-r}(Z_0)^\wedge
\end{array}
$$

and φ_4 is induced by the augmented cup, the others by the cup.

Theorem 8.1. *All φ_i are isomorphisms.*

Proof. We proceed stepwise.

φ_1 is an isomorphism by Tate's theorem.
φ_2 by a semilocal analysis and again by Tate's theorem.
φ_3 by the 5-lemma and the result for φ_1 and φ_2.
φ_4 by the augmented cup duality already done.
φ_5 by the 5-lemma and the result for φ_3 and φ_4.

So that's it.

Corollary 8.2. $H^1(G_{K/k}, C_{a,K}) = 0$.

Proof. It is dual to $H^1(Z_\alpha)$ which is 0 since we assumed the existence of a rational cycle of degree 1.

In the case of a curve, if we had worked with the true ideles J_K instead of our truncated ones I_K, we would also have obtained (essentially in the same way) the above corollary. Thus from the sequence

$$0 \to K(V)^* \to J_{a,K} \to C_{a,K} \to 0$$

we would get exactly

$$0 \to H^2(K(V)^*) \to H^2(J_{a,K})$$

thus recovering the fact that an element of the Brauer group which splits locally everywhere splits globally (H^2 is taken with $G_{K/k}$).

Furthermore, the curves exhibit one more duality, a self duality, of our group $F_{\mathfrak{p}}$. This is a local question. We take k a $\mathfrak{p}$-adic field, K a finite extension, Galois with group $G_{K/k}$, and consider the power series $k((t))$ and $K((t))$. We let F be our local group

$$F = K((t))^*/(1 + \mathfrak{m})$$

where $\mathfrak{m}$ is the maximal ideal. Then $1 + \mathfrak{m}$ is uniquely divisible, and so its cohomology is trivial. Hence

$$\mathbf{H}^r(G_{K/k}, K((t))^*) = \mathbf{H}^r(G_{K/k}, F).$$

We have the exact sequence

$$0 \to K^* \to F \to \mathbf{Z} \to 0$$

and a pairing
$$K((t))^* \times K((t))^* \to K^*$$
defined by

$$(f,g) = (-1)^{\mathrm{ord}f\ \mathrm{ord}g}(f^{\mathrm{ord}g}/g^{\mathrm{ord}f})(0),$$

which induces a pairing

$$F \times F \to K^*.$$

Now we get the commutative diagram

$$
\begin{array}{ccccccccc}
0 & \longrightarrow & \mathbf{H}^r(K^*) & \longrightarrow & \mathbf{H}^r(F) & \longrightarrow & \mathbf{H}^r(\mathbf{Z}) & \longrightarrow & 0 \\
 & & \downarrow & & \downarrow & & \downarrow & & \\
0 & \longrightarrow & \mathbf{H}^{2-r}(\mathbf{Z}^\wedge) & \longrightarrow & \mathbf{H}^{2-r}(F)^\wedge & \longrightarrow & \mathbf{H}^{2-r}(K^\wedge) & \longrightarrow & 0
\end{array}
$$

and by the five lemma, together with Tate's theorem, we see that the middle arrow is an isomorphism. Hence

$$\mathbf{H}^r(F) \quad \text{is dual to} \quad \mathbf{H}^{2-r}(F)$$

by the cup product.

Bibliography

[ArT 67] E. ARTIN and J. TATE, *Class Field Theory*, Benjamin 1967; Addison Wesley, 1991

[CaE 56] H. CARTAN and S. EILENBERG, *Homological Algebra*, Princeton Univ. Press 1956

[Gr 59] A. GROTHENDIECK, Sur quelques points d'algèbre homologique, *Tohoku Math. J.* 9 (1957) pp. 119-221

[Ho 50a] G. HOCHSCHILD, Local class field thoery, *Ann.Math.* 51 No. 2 (1950) pp. 331-347

[Ho 50b] G. HOCHSCHILD, Note on Artin's reciprocity law, *Ann. Math.* 52 No. 3 (1950) pp. 694-701

[HoN 52] G. HOCHSCHILD and T. NAKAYAMA, Cohomology in class field theory, *Ann.Math.* 55 No. 2 (1952) pp. 348-366

[HoS 53] G. HOCHSCHILD and J.-P. SERRE, Cohomology of group extensions, *Trans. AMS* 74 (1953) pp. 110-134

[Ka 55a] Y. KAWADA, Class formations, *Duke Math. J.* 22 (1955) pp. 165-178

[Ka 55b] Y. KAWADA, Class formations III, *J. Math. Soc. Japan* 7 (1955) pp. 453-490

[Ka 63] Y. KAWADA, Cohomology of group extensions, *J. Fac. Sci. Univ. Tokyo* 9 (1963) pp. 417-431

[Ka 69]   Y. KAWADA, Class formations, *Proc. Symp. Pure Math.* **20** AMS, 1969

[KaS 56] Y. KAWADA and I. SATAKE, Class formations II, *J. Fac. Sci. Univ. Tokyo* **7** (1956) pp. 353-389

[KaT 55] Y. KAWADA and J. TATE, On the Galois cohomology of unramified extensions of function fields in one variable, *Am. J. Math.* **77** No. 2 (1955) pp. 197-217

[La 57]   S. LANG, Divisors and endomorphisms on abelian varieties, *Amer. J. Math.* **80** No. 3 (1958) pp. 761-777

[La 59]   S. LANG, *Abelian Varieties*, Interscience, 1959; Springer Verlag, 1983

[La 66]   S. LANG, *Rapport sur la cohomologie des groupes*, Benjamin 1966

[La 71/93] S. LANG, *Algebra*, Addison-Wesley 1971, 3rd edn. 1993

[Mi 86]   J. MILNE, *Arithmetic Duality Theorems*, Academic Press, Boston, 1986

[Na 36]   T. NAKAYAMA, Über die Beziehungen zwischen den Faktorensystemen und der Normklassengruppe eines galoisschen Erweiterungskörpers, *Math. Ann.* **112** (1936) pp. 85-91

[Na 43]   T. NAKAYAMA, A theorem on the norm group of a finite extension field, *Jap. J. Math.* **18** (1943) pp. 877-885

[Na 41]   T. NAKAYAMA, Factor system approach ot the isomorphism and reciprocity theorems, *J. Math. Soc. Japan* **3** No. 1 (1941) pp. 52-58

[Na 52]   T. NAKAYAMA, Idele class factor sets and class field theory, *Ann. Math.* **55** No. 1 (1952) pp. 73-84

[Na 53]   T. NAKAYAMA, Note on 3-factor sets, *Kodai Math. Rep.* **3** (1949) pp. 11-14

[Se 73/94] J.-P. SERRE, *Cohomologie Galoisienne*, Benjamin 1973, Fifth edition, Lecture Notes in Mathematics No. 5, Springer Verlag 1994

[Sh 46]   I. SHAFAREVICH, On Galois groups of p-adic fields, *Dokl. Akad. Nauk SSSR* **53** No. 1 (1946) pp. 15-16 (see also *Collected Papers*, Springer Verlag 1989, p. 5)

[Ta 52] J. TATE, The higher dimensional cohomology groups of class field theory, *Ann. Math.* **56** No. 2 (1952) pp. 294-27

[Ta 62] J. TATE, Duality theorems in Galois cohomology over number fields, *Proc. Int. Congress Math. Stockholm* (1962) pp. 288-295

[Ta 66] J. TATE, The cohomology groups of tori in finite Galois extensions of number fields, *Nagoya Math. J.* **27** (1966) pp. 709-719

[We 51] A. WEIL, Sur la théorie du corps de classe, *J. Math. Soc. Japan* **3** (1951) pp. 1-35

Complementary References

A. ADEM and R.J. MILGRAM, *Cohomology of Finite Groups*, Springer-Verlag 1994

K. BROWN, *Cohomology of Groups*, Springer-Verlag 1982

S. LANG, *Algebraic Number Theory*, Addison-Wesley 1970; Springer-Verlag 1986, 2nd edn. 1994

S. MAC LANE, *Homology*, Springer-Verlag 1963, 4th printing 1995

Table of Notation

A_{p^∞} : Elements of A annihilated by a power of p

A_φ : If φ is a homomorphism, kernel of φ in A

$\hat{A}$: $\mathrm{Hom}(A, \mathbf{Q}/\mathbf{Z})$

A^G : Elements of A fixed by G

A_m : Kernel of the homomorphism $m_A :: A \to A$ such that $a \mapsto ma$

cd : Cohomological dimension

$\mathbf{F}_p$: $\mathbf{Z}/p\mathbf{Z}$

$\hat{G}$: Character group, $\mathrm{Hom}(G, \mathbf{Q}/\mathbf{Z})$

$\varkappa_G$: Natural homomorphism of A^G onto $H^0(G, A)$ or $\mathbf{H}^0(G, A)$

$\maltese_G$: Natural homomorphism of A_S into $\mathbf{H}^{-1}(G, A)$

$\mathrm{Galm}(G)$: Galois modules

$\mathrm{Galm}_p(G)$: Galois modules whose elements are annihilated by a p-power

$\mathrm{Galm}_{\mathrm{tor}}(G)$: Torsion Galois modules

G^c : Commutator group, or closure of commutator if G is topological

G_p : p-Sylow subgroup of G

Grab : Category of abelian groups

$h_{1/2}$: Herbrand quotient, order of H^2 divided by order of H^1

H_G : Functor such that $H_G(A) = A^G$

$\mathbf{H}_G$: Functor such that $\mathbf{H}_G(A) = A^G/\mathbf{S}_G A$

I_G : Augmentation ideal, generated by the elements $\sigma - e, \sigma \in G$

$M_G(A)$: Functions (sometimes continuous) from G into A

$\mathbf{M}_G$: $\mathbf{Z}[G] \otimes A$

M_G^S : Induced functions

$\operatorname{Mod}(G)$: Abelian category of G-modules

$\operatorname{Mod}(\mathbf{Z})$: Abelian category of abelian groups

scd : Strict cohomological dimension

$\mathbf{S}_G$: The relative trace, from a subgroup U of finite index, to G

$\mathbf{S}_G$: The trace, for a finite group G

Tr : Transfer of group theory

tr : Transfer of cohomology

$\mathbf{Z}[G]$: Group ring

Index

General Remarks

Lecture Notes are printed by photo-offset from the master-copy delivered in camera-ready form by the authors. For this purpose Springer-Verlag provides technical instructions for the preparation of manuscripts.

Careful preparation of manuscripts will help keep production time short and ensure a satisfactory appearance of the finished book. The actual production of a Lecture Notes volume normally takes approximately 8 weeks.

Authors receive 50 free copies of their book. No royalty is paid on Lecture Notes volumes.

Authors are entitled to purchase further copies of their book and other Springer mathematics books for their personal use, at a discount of 33,3 % directly from Springer-Verlag.

Commitment to publish is made by letter of intent rather than by signing a formal contract. Springer-Verlag secures the copyright for each volume.

Addresses:

Professor A. Dold
Mathematisches Institut
Universität Heidelberg
Im Neuenheimer Feld 288
D-69120 Heidelberg, Germany

Professor F. Takens
Mathematisch Instituut
Rijksuniversiteit Groningen
Postbus 800
NL-9700 AV Groningen
The Netherlands

Professor Bernard Teissier
École Normale Supérieure
45, rue d'Ulm
F-7500 Paris, France

Springer-Verlag, Mathematics Editorial
Tiergartenstr. 17
D-69121 Heidelberg, Germany
Tel.: *49 (6221) 487-410